Lecture Notes in Mathematics

Edited by A. Dold and B. Eckmann

786

Ivor J. Maddox

T0224602

Infinite Matrices of Operators

Springer-Verlag
Berlin Heidelberg New York 1980

Author

Ivor J. Maddox
Department of Pure Mathematics
Queen's University
Belfast BT7 1NN
United Kingdom

AMS Subject Classifications (1980): 40-02, 40C05, 40E05, 40F05, 40G05, 40H05, 46-02

ISBN 3-540-09764-3 Springer-Verlag Berlin Heidelberg New York
ISBN 0-387-09764-3 Springer-Verlag New York Heidelberg Berlin

Library of Congress Cataloging in Publication Data. Maddox, Ivor John. Infinite matrices of operators. (Lecture notes in mathematics ; 786) Bibliography: p. Includes index.
1. Operator theory. 2. Matrices, Infinite. 3. Summability theory. I. Title. II. Series: Lecture notes in mathematics (Berlin) ; 786.
QA3.L28 no. 786 [QA329] 510s [515.7'24] 80-11702

Printed in Germany

Printing and binding: Beltz Offsetdruck, Hemsbach/Bergstr.
2141/3140-543210

CONTENTS

1. Introduction

The classical theory of the transformation of complex sequences by complex infinite matrices is associated largely with the names of Toeplitz, Kojima, and Schur. The basic results of this theory may be conveniently found in the books by Hardy [19], Cooke [9], Maddox [40].

In Hardy's book one also finds detailed accounts of numerous special matrices, or means, e.g. the means of Cesàro, Nörlund, and Borel. Particular attention is given by Hardy to theorms of inclusion and consistency, as well as to theorems of Mercerian and Tauberian type.

Apart from the basic Toeplitz-Kojima-Schur theorems, Cooke, unlike Hardy, tends to deal with some of the more general aspects of the theory of infinite matrices, though like Hardy his treatment is essentially classical. Non-functional analytic methods are employed, and the sequences and matrices considered are restricted to be real or complex.

A decisive break with the classical approach was made by Abraham Robinson [66] in 1950, when he considered the action of infinite matrices of linear operators from a Banach space on sequences of elements of that space.

Our object in the present work is to give an account of some of the main developments which have occurred since Robinson's paper of 1950.

Most of our notation and terminology will be described in Section 2.

In the classical theory of matrix transformations, one of the basic problems is the characterization of matrices which map a sequence space (or merely a set of sequences) E into a sequence space (or set of sequences) F. The first step in this characterization is the determination of the Köthe-Toeplitz dual of E, also called the β-dual of E, where

$$E^{\beta} = \{a \in s : \sum_{k=1}^{\infty} a_k x_k \text{ converges for all } x \in E\}.$$

As usual, s denotes the linear space of all infinite sequences $a = (a_k)$ of complex numbers a_k.

The idea of dual sequence spaces was introduced by Köthe and Toeplitz [28], whose main results concerned α-duals; the α-dual of $E \subset s$ being defined as

$$E^{\alpha} = \{a \in s : \sum_{k=1}^{\infty} |a_k x_k| < \infty \text{ for all } x \in E\}.$$

An account of the theory of α-duals in the scalar case may be found in Köthe [27].

Another dual, the γ-dual, is defined by

$$E^{\gamma} = \{a \in s : \sup_n \left|\sum_{k=1}^{n} a_k x_k\right| < \infty \text{ for all } x \in E\}.$$

Certain topologies on a sequence space, involving β- and α-duality have been examined by Garling [15].

For certain special sequence spaces there are some interesting results given by Lascarides [34].

In Section 3 we investigate several generalized Köthe-Toeplitz duals which arise when the complex sequence (a_k) is replaced by a sequence (A_k) of linear operators. Thus, if X, Y are Banach spaces, each A_k is a linear operator on X into Y, and E is a nonempty set of sequences $x = (x_k)$, with $x_k \in X$, then we define

$$E^{\beta} = \{(A_k) : \sum_{k=1}^{\infty} A_k x_k \text{ converges in the Y-norm, for all } x \in E\}.$$

Section 4 is devoted to the characterization of a number of classes of matrix transformations of linear operators. Inter alia, one finds operator analogues of the theorems of Toeplitz, Kojima, Schur, and of the recent theorem of Crone [11] on infinite scalar matrices which map the Hilbert space ℓ_2 into itself.

In Sections 5, and 6 there is a discussion of Tauberian theorems, and the famous bounded consistency theorem of Mazur-Orlicz-Brudno. Section 7 introduces a new concept of operator Nörlund means and gives some results on the consistency of certain classes of these means.

2. Notation and terminology

By N,R,C we denote the natural, real, and complex numbers, respectively. Some frequently occurring sequence spaces are:

- s, the linear space of complex sequences
- ℓ_o, the space of finite complex sequences,
- c_o, the space of null complex sequences,
- c, the space of convergent complex sequences,
- [f], the space of strongly almost convergent complex sequences,
- f, the space of almost convergent complex sequences,
- ℓ_∞, the space of bounded complex sequences,
- ℓ_p, the space of p-absolutely summable complex sequences, where $0 < p < \infty$,
- w_p, the space of strongly Cesàro summable complex sequences of order 1 and index p, where $0 < p < \infty$.

Of the above spaces, only [f], f and w_p are not perhaps as standard as the others.

The space f was introduced by Lorentz [36]. We say that $(x_k) \in f$ if and only if there exists $\ell \in C$ such that

$$\frac{1}{r} \sum_{i=p+1}^{p+r} x_i \to \ell \ (r \to \infty, \text{ uniformly in } p \geq 0).$$

The space [f] was defined by Maddox [48]. We say that $(x_k) \in [f]$ if and only if there exists $\ell \in C$ such that

$$\frac{1}{r} \sum_{i=p+1}^{p+r} |x_i - \ell| \to 0 \ (r \to \infty, \text{ uniformly in } p \geq 0).$$

We have $c \subset [f] \subset f \subset \ell_\infty$ with strict inclusions, and c,[f], f are closed subspaces of ℓ_∞, which is a Banach space with $||x|| = \sup|x_k|$ for each $x = (x_k) \in \ell_\infty$.

The space w_p has been considered in [39] and [40]. We say that $(x_k) \in w_p$ if and only if there exists $\ell \in C$ such that

$$\frac{1}{n}\sum_{k=1}^{n}|x_k-\ell|^p \to 0 \ (n\to\infty).$$

If $(X,||.||)$ is any Banach space over C then we may define $c(X)$, the convergent X-valued sequences; $f(X)$, the almost convergent X-valued sequences, etc. Thus, e.g. $x = (x_k) \in \ell_\infty(X)$, where $x_k \in X$ for $k \in N$, if $\sup||x_k|| < \infty$. Consequently $\ell_\infty(X)$ becomes a Banach space, with the natural coordinatewise operations, and

$$||x|| = \sup||x_k||, \text{ for } x \in \ell_\infty(X).$$

Similarly, $x = (x_k) \in w_p(X)$, $0 < p < \infty$, if and only if there exists $\ell \in X$ such that

$$\frac{1}{n}\sum_{k=1}^{n}||x_k-\ell||^p \to 0 \ (n\to\infty).$$

Every space of complex sequences listed above may be generalized to a space of X-valued sequences merely by replacing the modulus in C by the norm in X, when appropriate.

If X, Y are Banach spaces then we denote by

$$B(X,Y)$$

the Banach algebra of bounded linear operators on X into Y, with the usual operator norm. Thus, if $T \in B(X,Y)$ the operator norm of T is

$$||T|| = \sup\{||Tx|| : x \in S\},$$

where $S = \{x \in X : ||x|| \leq 1\}$ is the closed unit sphere in X.

By U we mean the set of all $x \in X$ such that $||x|| = 1$. The zero element of X, and Y, is denoted by Θ.

The continuous dual of Y, i.e. the space of continuous linear functionals on Y, is $B(Y,C)$, and is written as Y^*. If $f \in Y^*$ and $y \in Y$ we use the notation

$$(f,y) = f(y).$$

For each $T \in B(X,Y)$ we denote the adjoint of T by T^*, where T^* is defined by

$$(f,Tx) = (T^*f,x), \text{ for all } f \in Y^* \text{ and all } x \in X.$$

We shall also write

$$S^* = \{f \in Y^* : ||f|| \leq 1\},$$

and make use of the well-known fact that, by the Hahn-Banach extension theorem, for every $y \in Y$ there exists $f \in S^*$ such that $||y|| = f(y)$.

The following concept was introduced by Robinson [66] and was termed the group norm by Lorentz and Macphail [37].

2.1 Definition. Let $(T_k) = (T_1, T_2, \ldots)$ be a sequence in $B(X,Y)$. Then the group norm of (T_k) is

$$||(T_k)|| = \sup||\sum_{k=1}^{n} T_k x_k||$$

where the supremum is over all $n \in N$ and all $x_k \in S$.

It may happen that the group norm is not finite, though we are usually concerned with problems which give rise to finite group norms.

2.2 Summation convention. A sum Σx_k without limits will always be over $k \in N$, i.e.

$$\Sigma x_k = \sum_{k=1}^{\infty} x_k.$$

Some elementary properties of group norms are given in:

2.3 Proposition. (i) If $(A_k) \in s(C^*)$ then the A_k may be identified with complex numbers a_k and

$$||(A_k)|| = \Sigma|a_k|,$$

whence the group norm is finite if and only if $a \in \ell_1$.

(ii) If (T_k) is a sequence in $B(X,Y)$ and we write

$$R_n = (T_n, T_{n+1}, T_{n+2}, \ldots)$$

then

(a) $||T_m|| \le ||R_n||$ *for all* $m \ge n$,

(b) $||R_{n+1}|| \le ||R_n||$ *for all* $n \in N$,

(c) $$||\sum_{k=n}^{n+p} T_k x_k|| \le ||R_n||.\ \max\{||x_k|| : n \le k \le n+p\},$$

for any x_k *and all* $n \in N$, *and all non-negative integers* p.

(iii) *If* (T_k) *is a sequence in* $B(X,Y)$ *then* $\Sigma||T_k|| < \infty$ *implies* $||(T_k)|| < \infty$. *Also*, $||(T_k)|| < \infty$ *implies* $\sup_k ||T_k|| < \infty$.

(iv) *If* Z *denotes the set of all sequences* $T = (T_k)$ *such that each group norm* $||T||$ *is finite then* Z *becomes a Banach space, with the natural operations, under the norm* $||T||$.

Proof. (i) There exist complex numbers a_k such that $A_k z = a_k z$ for all $z \in C$. Now for all $n \in N$ and all $x_k \in S$,

$$|\sum_{k=1}^{n} a_k x_k| \le \sum_{k=1}^{n} |a_k|.$$

Also, if $n \in N$, and we define $\text{sgn } z = |z|/z$ $(z \ne 0)$, $\text{sgn } 0 = 1$, then

$$|\sum_{k=1}^{n} a_k z_k| = \sum_{k=1}^{n} |a_k|$$

for $z_k = \text{sgn } a_k$ $(1 \le k \le n)$. It follows that $||(A_k)|| = \Sigma|a_k|$, with the understanding that the group norm is not finite when $\Sigma|a_k|$ diverges.

(ii) Let $x \in S$ and define $x_m = x$, $x_k = \theta$ $(n \le k < m)$. Then

$$||T_m x|| = ||\sum_{k=n}^{m} T_k x_k|| \le ||R_n||,$$

which yields (a).

Now take $x_k \in S$ for $n+1 \le k \le m$, so that

$$||\sum_{k=n+1}^{m} T_k x_k|| = ||T_n\Theta + \sum_{k=n+1}^{m} T_k x_k|| \le ||R_n||,$$

which yields (b).

Let M denote the max in (c). The case M = 0 is trivial. If M > 0 then x_k/M is in S for $n \le k \le n + p$, and (c) follows.

(iii) If $\Sigma||T_k|| < \infty$ and $n \in N$, $x_k \in S$, then

$$||\sum_{k=1}^{n} T_k x_k|| \le \sum_{k=1}^{n} ||T_k||\ ||x_k|| \le \Sigma||T_k||$$

whence $||(T_k)|| \le \Sigma||T_k||$. We note that the converse implication is generally false. For example, define $T_k \in B(\ell_\infty, \ell_\infty)$ by

$$T_k x = (0, 0, \ldots, x_1, 0, 0, \ldots)$$

with x_1 in the k-position, where $x = (x_k) \in \ell_\infty$. Then it is clear that $||T_k|| = 1$ for all $k \in N$, so $\Sigma||T_k||$ diverges. However, if

$$x^{(k)} = (x_1^{(k)}, x_2^{(k)}, \ldots) \in S$$

for $k \in N$ then $|x_k^{(n)}| \le ||x^{(n)}|| \le 1$ for all n and k, and so for any $n \in N$,

$$||\sum_{k=1}^{n} T_k x^{(k)}|| = ||(x_1^{(1)}, x_1^{(2)}, \ldots, x_1^{(n)}, 0, 0, \ldots)|| \le 1$$

whence $||(T_k)|| \le 1$. Moreover, on choosing $x^{(k)} = (1, 1, 1, \ldots)$ for $k \in N$ we see that $||(T_k)|| = 1$.

Now suppose that $||(T_k)|| < \infty$. By (ii)(a) above we have $||T_m|| \le ||R_1|| = ||(T_k)||$ for all $m \in N$, whence $\sup_m ||T_m|| \le ||(T_k)||$. The converse implication is always false in B(X,X), where X is a non-trivial Banach space, since we may take T_k as the identity operator for every k.

(iv) With the natural operations $T + T' = (T_k + T_k')$ and $\lambda T = (\lambda T_k)$, for

$\lambda \in C$, it is routine to check completeness. The proof uses the fact that $B(X,Y)$ is a Banach space with the usual norm.

2.4 Definition (Generalized Köthe-Toeplitz duals). *Let* X *and* Y *be Banach spaces and* (A_k) *a sequence of linear, but not necessarily bounded, operators* A_k *on* X *into* Y. *Suppose* E *is a nonempty subset of* $s(X)$. *Then the* α*-dual of* E *is defined as*

$$E^{\alpha} = \{(A_k) : \Sigma\|A_k x_k\| \text{ converges for all } (x_k) \in E\}.$$

The β*-dual of* E *is defined as*

$$E^{\beta} = \{(A_k) : \Sigma A_k x_k \text{ converges for all } (x_k) \in E\}.$$

We remark that convergence is in the norm of Y, in the definition of E^{β}.

In case $X = Y = C$ and the A_k are identified with complex numbers a_k, then $E \subset s$ and

$$E^{\alpha} = \{a : \Sigma|a_k x_k| < \infty \text{ for all } (x_k) \in E\},$$

$$E^{\beta} = \{a : \Sigma a_k x_k \text{ converges for all } (x_k) \in E\}.$$

The α and β-duals of the commonly occuring sequence spaces are all well-known, e.g. $c_0^{\alpha} = c_0^{\beta} = c^{\alpha} = c^{\beta} = \ell_{\infty}^{\alpha} = \ell_{\infty}^{\beta} = \ell_1$.

2.5 Definition. *Let* X *and* Y *be Banach spaces and* $A = (A_{nk})$ *an infinite matrix of linear, but not necessarily bounded, operators* A_{nk} *on* X *into* Y.

Suppose E *is a nonempty subset of* $s(X)$ *and* F *is a nonempty subset of* $s(Y)$. *Then we define the matrix class* (E,F) *by saying that* $A \in (E,F)$ *if and only if, for every* $x = (x_k) \in E$,

$$A_n(x) = \Sigma A_{nk} x_k = \sum_{k=1}^{\infty} A_{nk} x_k$$

converges in the norm of Y, *for each* n, *and the sequence*

$$Ax = (\Sigma A_{nk}x_k)_{n\in N}$$

belongs to F.

In case X = Y = C, and the A_{nk} are identified with complex numbers a_{nk} we shall make use of the following conditions in order to characterize some of the important matrix classes. It is to be understood that a condition such as (2.1) involves the convergence of $\Sigma|a_{nk}|$ for each n. As usual, a summation without limits is over $k \in N$. unless otherwise indicated.

(2.1) $$\sup_n \Sigma|a_{nk}| < \infty,$$

(2.2) $$\sup_n \Sigma|\Delta a_{nk}| < \infty, \text{ where } \Delta a_{nk} = a_{nk} - a_{n,k+1},$$

(2.3) $$\Sigma|a_{nk}| \text{ converges uniformly in } n,$$

(2.4) $$\sup_{n,k}|a_{nk}| < \infty,$$

(2.5) $$\lim_n \Sigma|a_{nk}| = 0,$$

(2.6) $$\sup_k \sum_{n=1}^{\infty} |a_{nk}|^p < \infty, \text{ where } p \geq 1,$$

(2.7) $$\lim_n a_{nk} \text{ exists for each } k,$$

(2.8) $$\lim_n a_{nk} = 0 \text{ for each } k,$$

(2.9) $$\lim_n \Sigma a_{nk} \text{ exists},$$

(2.10) $$\lim_n \Sigma a_{nk} = 1$$

(2.11) $$\sup_n \sum_{r=0}^{\infty} 2^{r/p} \max\{|a_{nk}| : 2^r \leq k < 2^{r+1}\} < \infty,$$

where $0 < p < 1$,

(2.12) $$\sup_n \sum_{r=0}^{\infty} 2^{r/p} (\Sigma_r |a_{nk}|^q)^{1/q} < \infty,$$

where $p \geq 1$, $1/p + 1/q = 1$, and Σ_r is over $2^r \leq k < 2^{r+1}$.

If $p = 1$ in (2.12) we interpret Σ_r as $\max\{|a_{nk}| : 2^r \leq k < 2^{r+1}\}$.

2.6 Theorem. $(\ell_\infty,\ell_\infty) = (c,\ell_\infty) = (c_0,\ell_\infty)$, and $A \in (\ell_\infty,\ell_\infty)$ if and only if (2.1) holds.

A proof, along classical lines, may be found in Petersen [62].

2.7 Theorem (KOJIMA-SCHUR). $A \in (c,c)$ if and only if (2.1), (2.7), (2.9) hold.

See Schur [68], or Hardy [19], or Cooke [9].

2.8 Definition. If $A \in (c,c)$ we say that A is conservative. The characteristic of a conservative A is defined to be

$$\chi(A) = \lim_n \Sigma a_{nk} - \Sigma(\lim_n a_{nk}).$$

If $\chi(A) = 0$, we say that A is conull, whilst if $\chi(A) \neq 0$, we say that A is coregular.

2.9 Theorem (TOEPLITZ). $A \in (c,c)$, leaving the limit of every convergent sequence invariant, if and only if (2.1), (2.8), (2.10) hold.

See Toeplitz [76], or [9], [19], [40].

2.10 Theorem. $A \in (\gamma,c)$, where $\gamma = \{x : \Sigma x_k$ converges$\}$, if and only if (2.2), (2.7) hold.

See Cooke [9].

2.11 Theorem (SCHUR). $A \in (\ell_\infty,c)$ if and only if (2.3), (2.7) hold.

See Schur [68], or Maddox [40], p.169. Also, one sees from the proof in Maddox [40], p.169 that $A \in (\ell_\infty, c)$ if and only if (2.1), (2.7) and

(2.13) $$\lim_n \Sigma |a_{nk} - \lim_n a_{nk}| = 0.$$

We remark that another set of necessary and sufficient conditions for $A \in (\ell_\infty, c)$ is (2.7) and

$$\lim_n \Sigma |a_{nk}| = \Sigma |\lim_n a_{nk}|.$$

2.12 Theorem. $A \in (\ell_\infty, c_0)$ *if and only if (2.5) holds.*

See, for example, Maddox [40], p.169. An interesting consequence of Theorem 2.11 is that strong and weak convergence of sequences coincide in ℓ_1 (Maddox [40], p.170).

2.13 Theorem.

(i) $A \in (\ell_1, \ell_\infty)$ *if and only if (2.4) holds.*

(ii) $A \in (\ell_1, \ell_p)$, *where* $p \geq 1$, *if and only if (2.6) holds.*

See Hahn [16] for (i) of Theorem 2.13, and Maddox [40], p.167 for (ii). The condition for $A \in (\ell_1, \ell_1)$ was first given by Knopp and Lorentz [26].

2.14 Theorem. $A \in (\ell_\infty, \ell_1)$ *if and only if*

(2.14) $$\sup \sum_k \Big| \sum_{n \in E} a_{nk} \Big| < \infty,$$

where the supremum is taken over all finite subsets E *of* N.

See Zeller [82], Mehdi [54].

Some remarks on Theorem 2.14 may be of interest. The condition

(2.15) $$\sum_n \sum_k |a_{nk}| < \infty$$

is obviously sufficient for $A \in (\ell_\infty, \ell_1)$, since $x \in \ell_\infty$ implies

$$\sum_n \left|\sum_k a_{nk} x_k\right| \le ||x|| \sum_n \sum_k |a_{nk}|.$$

But (2.15) is not necessary. For ℓ_1 is infinite dimensional, so by the Dvoretzky-Rogers theorem [14], there is a series Σa_n in ℓ_1 such that $\Sigma ||a_n|| = \infty$ but Σa_n is unconditionally convergent, where $a_n = (a_{nk}) = (a_{n1}, a_{n2}, \ldots)$. Now Σa_n unconditionally convergent implies

$$\sup_E \left|\left| \sum_{n \in E} a_n \right|\right| < \infty, \tag{2.16}$$

where E denotes a finite subset of N. But (2.16) is exactly the condition (2.14), so $A = (a_{nk})$ is in (ℓ_∞, ℓ_1) by Theorem 2.14, but

$$\sum_n ||a_n|| = \sum_n \sum_k |a_{nk}| = \infty.$$

Since the Dvoretzky-Rogers Theorem is of some depth, it is worth observing that the result for the particular space ℓ_1 was dealt with by Macphail [38], who argues as follows. Let S be a finite sequence $(x_1, x_2, \ldots, x_n)$ of elements x_k from a Banach space X. Define

$$|S| = \sum_{k=1}^{p} ||x_k||,$$

$$|S|^* = \sup_E \left|\left| \sum_{k \in E} x_k \right|\right|,$$

where E is any subset of $\{1,2,\ldots,p\}$. If

$$\inf |S|^*/|S| = 0, \tag{2.17}$$

where the infimum is over all finite sequences S with $|S| > 0$, then there is an unconditionally convergent series from X which is not absolutely convergent

In case $X = \ell_1$ define

$$S_1 = ((-1,1,0,0,0,\ldots)),$$

$$S_2 = ((-1,-1,1,1,0,0,0,\ldots),$$
$$(-1,1,-1,1,0,0,0,\ldots)),$$

$$S_3 = ((-1,-1,-1,-1,1,1,1,1,0,0,0,\ldots),$$
$$(-1,-1,1,1,-1,-1,1,1,0,0,0,\ldots),$$
$$(-1,1,-1,1,-1,1,-1,1,0,0,0,\ldots)),$$

and so on. Then $|S_n| = n2^n$, and if we use the notation

$$S_n = (R_1^n, R_2^n, \ldots, R_n^n)$$

with $R_k^n(m)$ the m-th term of R_k^n then

$$|S|^* = \sup_E \|\sum_{k\in E} R_k^n\|$$

$$= \sup_E \sum_{m=1}^{\infty} |\sum_{k\in E} R_k^n(m)|$$

$$= \sup_E 2^n \int_0^1 |\sum_{k\in E} r_k(t)|dt$$

$$\leq \sup_E 2^n \{\int_0^1 (\sum_{k\in E} r_k(t))^2 dt\}^{1/2}$$

$$= \sup_E 2^n |E|^{1/2} \leq 2^n . n^{1/2}.$$

In the above, r_k denotes a Rademacher function, such that r_k is defined on $[0,1]$ and $r_k(t)$ takes the values -1 and 1 alternately on open intervals of length 2^{-k}. At the end-points of the intervals $r_k(t)$ is defined to be 0. It is well-known that $\{r_n\}$ is orthonormal in $L_2[0,1]$, a fact that was used above. Also in the above $|E|$ denotes the number of members of E.

It follows that $|s_n|^*/|s_n| \le n^{-1/2}$ which implies (2.17).

2.15 Theorem.

(i) *If* $0 < p < 1$ *then* $A \in (w_p, c)$ *if and only if (2.7) and (2.11) hold.*

(ii) *If* $p \ge 1$ *then* $A \in (w_p, c)$ *if and only if (2.7), (2.9) and (2.12) hold.*

See Maddox [39], [40].

The following consequence of Theorem 2.15(i) is important in connection with Kuttner's theorem [29], [39].

2.16 Corollary. *If* $0 < p < 1$ *then* $(w_p, c) \subset (\ell_\infty, c)$.

Proof. Take any n and $s \in N$, and any $m \ge 2^s$. Since $p < 1$,

$$\begin{aligned}
\sum_{k=m}^{\infty} |a_{nk}| &\le \sum_{r=s}^{\infty} \Sigma_r |a_{nk}| \\
&\le \sum_{r=s}^{\infty} 2^r \max\{|a_{nk}| : 2^r \le k < 2^{r+1}\} \\
&= \sum_{r=s}^{\infty} 2^{r/p} . 2^{r(p-1)/p} \max\{|a_{nk}| : 2^r \le k < 2^{r+1}\} \\
&\le 2^{s(p-1)/p} \sum_{r=0}^{\infty} 2^{r/p} \max\{|a_{nk}| : 2^r \le k < 2^{r+1}\}.
\end{aligned}$$

It follows from (2.11), since $p < 1$, that $\Sigma|a_{nk}|$ is uniformly convergent in n, and our result is a consequence of Theorem 2.11.

We now consider the current situation regarding the characterization of the matrices (a_{nk}) in (ℓ_r, ℓ_s) in terms of the a_{nk} alone. For arbitrary r, s with $1 \le r,s \le \infty$ the problem is unsolved. Some special cases are listed already (see Theorems 2.6, 2.13, 2.14). Even when $r = s$ only the cases $r = 1$, $r = \infty$ had been dealt with, until quite recently. Of particular interest was the case $r = s = 2$, relating to the Hilbert space ℓ_2, which had eluded many efforts.

In 1971, Crone [11] finally solved the problem for (ℓ_2,ℓ_2) with the following result.

2.17 Theorem (CRONE). $A \in (\ell_2,\ell_2)$ if and only if

(i) the rows of A are in ℓ_2,

(ii) $(A^*A)^n$ is defined for all $n \in N$,

(iii) $\sup_n \sup_i |[(A^*A)^n]_{ii}|^{1/n} < \infty$.

By A^* we denote the conjugate transpose of A.

An operator version of Crone's theorem will be given in Section 4.

To my knowledge, only the cases listed below have been characterized in the sense of necessary and sufficient conditions in terms of the a_{nk} alone:

(ℓ_r,ℓ_1) for $1 \le r \le \infty$,

(ℓ_∞,ℓ_s) for $1 < s \le \infty$,

(ℓ_r,ℓ_∞) for $1 \le r < \infty$,

(ℓ_1,ℓ_s) for $1 < s < \infty$,

(ℓ_2,ℓ_2),

(ℓ_r,ℓ_s) for $0 < r \le 1$ and $r \le s < \infty$.

The last case in the list does not appear to have been published but was given in the 1970 Ph.D. thesis of a former student of mine (Dr.J.W. Roles). We state it for completeness:

2.18 Theorem. Let $0 < r \leq 1$ and $r \leq s < \infty$. Then $A \in (\ell_r, \ell_s)$ if and only if

$$\sup_k \sum_{n=1}^{\infty} |a_{nk}|^s < \infty.$$

The reader is referred to the sources indicated for characterizations of the following matrix classes:

(i) (f,c); Lorentz [36], Stieglitz [70], [71].

(ii) (c,f); King [25], but see the remark of Stieglitz [70], p.19.

(iii) (ℓ_∞,f), (f,f); Stieglitz [70], Duran [13].

(iv) $([f],c)$, $([f],f)$, $(\ell_\infty,[f])$; Maddox [48], [49].

A useful survey on certain matrix classes is given by Stieglitz and Tietz [72].

We now establish notation and terminology connected with the summability of sequences by infinite matrices of operators. For simplicity we restrict attention to bounded linear operators.

2.19 Definition (Summability). Let X and Y be Banach spaces and $A = (A_{nk})$ an infinite matrix of operators $A_{nk} \in B(X,Y)$. Let $x = (x_k) \in s(X)$.

(i) We say that x is summable (or limitable) by A to ℓ if and only if there exists $\ell \in Y$ such that the series

$$A_n(x) = \Sigma A_{nk} x_k$$

converges in the norm of Y, for each n, and $A_n(x) \to \ell$ $(n \to \infty)$.

When x is summable by A to ℓ we write $x_n \to \ell(A)$, and $\ell = A\text{-}\lim x$.

(ii) If $B = (B_{nk})$ then we say that A implies B if and only if every sequence x which is summable by A is also summable by B, and A-lim x = B-lim x.

(iii) We say that A is regular (or a Toeplitz matrix) if and only if every convergent sequence x is also summable by A and lim x = A-lim x.

(iv) We define the summability field (A) of a matrix A to be

$$(A) = \{x \in s(X) : Ax = (A_n(x)) \in c(Y)\}.$$

Several authors use c_A instead of (A); see for example Wilansky [79].

We remark that not all summability methods of interest are given by matrices. For example, the important notion of strong summability is not given by a matrix. Recall, in connection with strong Cesàro summability of index $p > 0$, that $x \in s$ is said to be w_p summable to ℓ if and only if

$$\frac{1}{n} \sum_{k=1}^{n} |x_k - \ell|^p \to 0 \quad (n \to \infty).$$

Strong summability has been used in the theory of Fourier series in connection with extension of Fejér's classical result on (C,1) summability.

Let S[f] denote the Fourier series of a function f. It was shown by Hardy and Littlewood [20] that if $f \in L_2$ then S[f] is summable w_2 to f(x) at every continuity point x of f.

For $f \in L_1$ it was proved by Marcinkiewicz [51] that S[f] is summable w_2 to f(x) for almost every x. Generalizations of these results may be found in Zygmund [86], volume II.

Also, strong Cesàro summability has found application in ergodic theory; see for example the chapter on mixing in Halmos [17].

For reference we mention the well-known classical methods of summability associated with the names of Abel, Cesàro, and Nörlund. As is conventional with these methods the summations start at k = 0.

(i) We say that $x = (x_0, x_1, \ldots)$ is Abel summable to ℓ if and only if

$$\sum_{k=o}^{\infty} x_k y^k \text{ converges for } 0 < y < 1,$$

and

$$(1-y) \sum_{k=o}^{\infty} x_k y^k \to \ell \ (y \to 1-).$$

Abel summability is often defined for series Σa_k rather than sequences. Thus we say that Σa_k is Abel summable to a number s, and write $\Sigma a_k = s$ (Abel), if and only if

$$f(y) = \sum_{k=o}^{\infty} a_k y^k \text{ converges for } 0 < y < 1,$$

and $f(y) \to s \ (y \to 1-)$.

(ii) Let $\alpha > -1$. Define $A_o^\alpha = 1$, and for $n \in N$,

$$A_n^\alpha = (\alpha+1)(\alpha+2) \ldots (\alpha+n)/n!$$

Then x is said to Cesàro summable of order α, i.e. (C,α) summable, to ℓ, if and only if

$$\frac{1}{A_n^\alpha} \sum_{k=o}^{n} A_{n-k}^{\alpha-1} x_k \to \ell \ (n \to \infty).$$

(iii) Let $q = (q_0, q_1, \ldots)$ be a complex sequence such that

$$Q_n = q_0 + q_1 + \ldots + q_n$$

is nonzero for all $n \geq 0$. Then x is (N,q) summable to ℓ if and only if

$$\frac{1}{Q_n} \sum_{k=o}^{n} q_{n-k} x_k \to \ell \ (n \to \infty).$$

By choosing $q_n = A_n^{\alpha-1}$ in (iii) we see that the (C,α) mean is a special type of Nörlund mean (N,q).

3. Generalized Köthe - Toeplitz duals

We shall determine Köthe-Toeplitz duals, in the operator case, for the more interesting sequence spaces. The results indicate the gap between the operator and the ordinary scalar case. For example, in the scalar case it is well-known that

$$c_0^\beta = c^\beta = \ell_\infty^\beta = \ell_1.$$

However, for the operator case we shall see that it is possible only to assert that

$$\ell_\infty^\beta(X) \subset c^\beta(X) \subset c_0^\beta(X).$$

In the following propositions we suppose in general that (A_k) is a sequence of linear, but not necessarily bounded, operators A_k mapping a Banach space X into a Banach space Y.

3.1 Proposition. $(A_k) \in c_0^\beta(X)$ *if and only if there exists* $m \in N$ *such that* $A_k \in B(X,Y)$ *for all* $k \geq m$ *and*

$$||R_m|| = ||(A_m, A_{m+1}, \ldots)|| < \infty.$$

Proof. For the sufficiency, let $(x_k) \in c_0(X)$ and take $n \geq m$. By Proposition 2.3,

$$||\sum_{k=n}^{n+p} A_k x_k|| \leq ||R_m||. \max\{||x_k|| : n \leq k \leq n+p\},$$

whence the completeness of Y implies that $\Sigma A_k x_k$ converges.

Conversely, suppose $(A_k) \in c_0^\beta(X)$ but that no m exists for which $A_k \in B(X,Y)$ for all $k \geq m$. Then there exists a strictly increasing sequence (k_i) of natural numbers and a sequence (z_i) in S such that

$$||A_{k_i} z_i|| > i, \text{ for } i \in N.$$

Define $x_k = z_i/i$ for $k = k_i$ and $x_k = \Theta$ otherwise. Then $(x_k) \in c_0(X)$ but $||A_k x_k|| > 1$ for infinitely many k, contrary to the fact that $\Sigma A_k x_k$ converges. Hence the A_k, are ultimately, bounded.

Now $c_0(X)$ is a Banach space under the norm $||x|| = \sup||x_k||$, with $x = (x_k) \in c_0(X)$. Since $\sum_{k=m}^{\infty} A_k x_k$ converges for all $x \in c_0(X)$ the Banach-Steinhaus theorem yields a constant H such that

$$||\sum_{k=m}^{\infty} A_k x_k|| \le H||x||$$

on $c_0(X)$. Now take any $n \in N$ and any $x_m, x_{m+1}, \ldots, x_{m+n}$ in S, and define

$$x = (\Theta,\Theta,\ldots,\Theta,x_m,\ldots,x_{m+n},\Theta,\Theta,\ldots).$$

Then $x \in c_0(X)$ and $||x|| \le 1$, whence $||R_m|| \le H$, which completes the proof.

3.2 <u>Proposition.</u> $(A_k) \in c^{\beta}(X)$ <u>if and only if</u> (i) <u>there exists</u> $m \in N$ <u>such that</u> $A_k \in B(X,Y)$ <u>for all</u> $k \ge m$, (ii) $||R_m|| < \infty$, <u>and</u> (iii) ΣA_k <u>converges in the strong operator topology</u>, i.e. $\Sigma A_k x$ <u>converges for each</u> $x \in X$.

<u>Proof.</u> To show that (iii) is necessary we take any $x \in X$. Then the sequence $(x,x,x,\ldots) \in c(X)$, and so $\Sigma A_k x$ converges. Also, since

$$c^{\beta}(X) \subset c_0^{\beta}(X) \tag{3.1}$$

the necessity of (i) and (ii) follows from the necessity part of Proposition 3.1.

For the sufficiency, let $x_k \to \ell (k \to \infty)$. Then $(x_k - \ell) \in c_0(X)$ and since $\Sigma A_k \ell$ converges, we see by the sufficiency part of Proposition 3.1 that $\Sigma A_k x_k$ converges with sum $\Sigma A_k \ell + \Sigma A_k (x_k - \ell)$.

We note that the inclusion (3.1) may be strict. For example, let $X = Y = \ell_\infty$ and define

$$A_k x = (0,0,\ldots,x_1,0,0,\ldots)$$

with x_1 in the k-position, where $x = (x_k) \in \ell_\infty$. Then $A_k \in B(X,Y)$ for all $k \in N$ and $||(A_k)|| = 1$, so that $(A_k) \in c_0^\beta(X)$. However, (iii) of Proposition 3.2 fails, since

$$s^{(n)} = \sum_{k=1}^{n} A_k(1,0,0,\ldots) = (1,1,\ldots,1,0,0,0,\ldots)$$

is such that $(s^{(n)})$ is not Cauchy in ℓ_∞.

3.3 Proposition. $(A_k) \in \ell_\infty^\beta(X)$ if and only if

(i) $(A_k) \in c_0^\beta(X)$, and (ii) $||R_n|| \to 0$ $(n \to \infty)$.

Proof. For the sufficiency, let $x = (x_k) \in \ell_\infty(X)$ with $||x|| = \sup||x_k||$. By (i) there exists m such that $||R_m|| < \infty$, so for $n \geq m$,

$$\left|\left| \sum_{k=n}^{n+p} A_k x_k \right|\right| \leq ||R_n||.||x||$$

whence $\Sigma A_k x_k$ converges by (ii).

The necessity of (i) is trivial. Now suppose, if possible, that (ii) fails, say

$$\lim\sup_n ||R_n|| = 3p > 0.$$

Then there exist natural numbers $n(1) \geq m(1) \geq m$ and $z_{m(1)},\ldots,z_{n(1)}$ in S such that

$$\left|\left| \sum_{m(1)}^{n(1)} A_k z_k \right|\right| > p.$$

Choose $m(2) > n(1)$ such that $||R_{m(2)}|| > 2p$. Then there exist $n(2) \geq m(2)$

and $z_{m(2)}, \ldots, z_{n(2)}$ in S such that

$$\left|\left| \sum_{m(2)}^{n(2)} A_k z_k \right|\right| > p.$$

Proceed in this way, and define $x_k = \Theta$ $(k < m(1))$, $x_k = z_k$ $(m(1) \le k \le n(1))$, $x_k = \Theta$ $(n(1) < k \le m(2))$, $x_k = z_k$ $(m(2) \le k \le n(2))$, etc. Then $||x|| = \sup ||x_k|| \le 1$ so $x \in \ell_\infty(X)$, but $\Sigma A_k x_k$ diverges, which gives a contradiction. This proves the proposition.

We note that the inclusion

$$\ell_\infty^\beta(X) \subset c^\beta(X)$$

may be strict. For example, let $X = Y = c_0$ and define

$$A_k x = (0,0,\ldots,x_k,0,0,\ldots)$$

with x_k in the k-position, where $x = (x_k) \in c_0$. Then $A_k \in B(X,Y)$ for all $k \in N$, and it is easy to see that $||R_n|| = 1$ for all $n \in N$ so that Proposition 3.3(ii) is not satisfied. However, for $x \in c_0$,

$$\left\{ \sum_{k=1}^{n} A_k x \right\} - x = (0,0,\ldots,-x_{n+1}, -x_{n+2}, \ldots)$$

and so ΣA_k converges, whence $(A_k) \in c^\beta(X)$ by Proposition 3.2.

Robinson [66], THEOREMS III, VI, first gave Proposition 3.2 although the case for unbounded operators was stated without proof. Moreover, Robinson makes no use of the Banach-Steinhaus theorem which is used implicity in our proof, since 3.2 depends on 3.1.

Proposition 3.3, for bounded operators, was given by Maddox [45], LEMMA 1.

It is worth observing that although $c_0^\alpha = c_0^\beta$ in the scalar case, we can only assert in general that $c_0^\alpha(X) \subset c_0^\beta(X)$, where the inclusion may be strict.

This follows from the next proposition, Proposition 3.1, and the remark made in Proposition 2.3(iii).

3.4 Proposition. $(A_k) \in c_0^\alpha(X)$ if and only if (i) there exists $m \in N$ such that $A_k \in B(X,Y)$ for all $k \geq m$ and

$$\text{(ii)} \qquad \sum_{k=m}^{\infty} ||A_k|| < \infty.$$

Proof. The sufficiency is trivial, and the necessity of (i) follows from the argument of Proposition 3.1. If we suppose (ii) fails then there exists a strictly increasing sequence $(n(i))$ and a sequence (z_k) in S such that $2||A_k z_k|| \geq ||A_k||$ and

$$\sum_{1+n(i)}^{n(i+1)} ||A_k|| > 2i, \text{ for } i \in N.$$

Define $x_k = z_k/i$ for $n(i) < k \leq n(i+1)$. Then $x \in c_0(X)$ and

$$\sum_{1+n(i)}^{n(i+1)} ||A_k x_k|| > 1,$$

contrary to the fact that $\Sigma||A_k x_k|| < \infty$. The proposition is now proved.

It is clear that the conditions of Proposition 3.4 are also necessary and sufficient for $(A_k) \in \ell_\infty^\beta(X)$ whence we have

$$c_0^\alpha(X) = c^\alpha(X) = \ell_\infty^\alpha(X),$$

which is the natural extension of the scalar case, where $c_0^\alpha = c^\alpha = \ell_\infty^\alpha$.

We shall have use for the next result, which is a special case of Proposition 3.3.

3.5 Proposition. Let $f_k \in X^*$ for $k \in N$. Then $(f_k) \in \ell_\infty^\beta(X)$ if and only if $\Sigma||f_k|| < \infty$.

Proof. The sufficiency is immediate. By Propositions 3.3 and 3.1 it is necessary that the group norm

$$H = \sup \left| \sum_{k=1}^{n} f_k(x_k) \right| < \infty.$$

Choose $x_k \in S$ such that $2|f_k(x_k)| \geq ||f_k||$, for $k \in N$. Take any $n \in N$ and define

$$\lambda_k = \operatorname{sgn} f_k(x_k) \text{ for } 1 \leq k \leq n,$$

$$\lambda_k = 0 \text{ for } k > n.$$

Then $(\lambda_k x_k)$ is a sequence in S, and so

$$\sum_{k=1}^{n} |f_k(x_k)| = \left| \sum_{k=1}^{n} \lambda_k f_k(x_k) \right| \leq H,$$

whence

$$\sum_{k=1}^{n} ||f_k|| \leq 2H,$$

which yields the result.

As an application of the above results we prove the following theorem of Thorp [74], THEOREM 3.2. We remark that Thorp uses notation different from ours.

3.6 Theorem. Let $A_k \in B(X,Y)$ for $k \in N$. Then

(3.2) $$\ell_1(B(X,Y)) \subset \ell_\infty^\beta(X),$$

and equality holds in (3.2) if and only if Y is finite dimensional.

Proof. If $(A_k) \in \ell_1(B(X))$ and $x \in \ell_\infty(X)$ then

$$\Sigma ||A_k x_k|| \leq ||x|| \Sigma ||A_k|| < \infty,$$

whence $\Sigma A_k x_k$ converges, which implies (3.2).

Now suppose Y has finite dimension n and that $(b_1, b_2, \ldots, b_n)$ is a Hamel base for Y. Then $y \in Y$ implies

$$y = \sum_{i=1}^{n} \lambda_i(y) b_i$$

where each $\lambda_i \in Y^*$. Hence, for $z \in X$ and $k \in N$,

$$A_k z = \sum_{i=1}^{n} \lambda_i(A_k z) b_i \tag{3.3}$$

with $\lambda_i \circ A_k \in X^*$. Now let $(A_k) \in \ell_\infty^\beta(X)$ so that

$$\Sigma(\lambda_i \circ A_k) x_k$$

converges for all $x \in \ell_\infty(X)$ and each i with $1 \le i \le n$.

By Proposition 3.5 it follows that

$$M_i = \Sigma ||\lambda_i \circ A_k|| < \infty$$

for each i.

Write $M = \max\{||b_i|| : 1 \le i \le n\}$. Then (3.3) implies

$$||A_k|| \le M \sum_{i=1}^{n} ||\lambda_i \circ A_k||,$$

and so

$$\Sigma||A_k|| \le M \sum_{i=1}^{n} M_i < \infty.$$

Hence equality holds in (3.2).

Now suppose Y is infinite dimensional. Then the Dvoretzky-Rogers theorem [14] implies the existence of a series Σy_k in Y which is unconditionally convergent but not absolutely convergent.

Take $f \in X^*$ with $||f|| = 1$ and define $A_k \in B(X,Y)$ for each $k \in N$ by

$$A_k z = f(z) y_k, \text{ for each } z \in X.$$

If $x \in \ell_\infty(X)$ then $(f(x_k)) \in \ell_\infty$ and so $\Sigma f(x_k) y_k$ converges since Σy_k is unconditionally convergent (see e.g. Jameson [23], pp. 342-3). Thus $\Sigma A_k x_k$ converges, whence $(A_k) \in \ell_\infty^\beta(X)$. But $||A_k|| = ||y_k||$ so that (A_k) is not in $\ell_1(B(X,Y))$, so that (3.2) is strict, and the proof is complete.

We next consider the duals of the $\ell_p(X)$ spaces.

3.7 <u>Proposition</u>. <u>Let</u> $0 < p \leq 1$. <u>Then</u> $(A_k) \in \ell_p^\beta(X)$ <u>if and only if there exists</u> $m \in N$ <u>such that</u> $A_k \in B(X,Y)$ <u>for all</u> $k \geq m$ <u>and</u> $\sup_{k\geq m}||A_k|| < \infty$.

<u>Proof</u>. If $\Sigma||x_k||^p < \infty$ and $H = \sup_{k\geq m}||A_k|| < \infty$, then by a familiar inequality (see e.g. Maddox [40], p.22),

$$\left(\sum_{k=m}^{\infty} ||A_k x_k||\right)^p \leq \sum_{k=m}^{\infty} ||A_k x_k||^p \leq \sum_{k=m}^{\infty} ||A_k||^p ||x_k||^p$$

$$\leq H^p \, \Sigma||x_k||^p.$$

Hence $\sum_{k=m}^{\infty} A_k x_k$ is absolutely convergent, and so $\Sigma A_k x_k$ is convergent. This proves the sufficiency.

Conversely, if $(A_k) \in \ell_p^\beta(X)$ and no such m exists, there is a sequence (k_i) and a sequence (z_i) in S such that

$$||A_{k_i} z_i|| > i^{2/p}, \text{ for } i \in N.$$

Define $x_k = z_i/i^{2/p}$ for $k = k_i$ and $x_k = \theta$ otherwise. Then $x \in \ell_p(X)$ but $||A_k x_k|| > 1$ for infinitely many k, contrary to the fact that $\Sigma A_k x_k$ converges.

Finally, assuming that $\sup_{k\geq m}||A_k|| = \infty$, an argument similar to that just given yields the existence of $w \in \ell_p(X)$ such that $\Sigma A_k w_k$ diverges.

We remark that the case $p = 1$ of Proposition 3.7, subject to the restriction that all the A_k are in $B(X,Y)$ was given by Thorp [74].

Also, we note that the argument of Proposition 3.7 shows that, for $0 < p \leq 1$, the β-dual of $\ell_p(X)$ is equal to its α-dual.

Subject to the restriction that all the A_k are in $B(X,Y)$, the following result was given by Thorp [74].

3.8 Proposition. *Let* $1 < p < \infty$. *Then* $(A_k) \in \ell_p^\beta(X)$ *if and only if there exists* $m \in N$ *such that* $A_k \in B(X,Y)$ *for all* $k \geq m$ *and*

$$\sup \sum_{k=m}^{\infty} ||A_k^* f||^q < \infty, \tag{3.4}$$

where q *is such that* $(1/p) + (1/q) = 1$, *and the supremum in (3.4) is taken over all* $f \in S^*$.

Proof. For the sufficiency, let $x \in \ell_p(X)$, let H denote the sup in (3.4) and take $r \geq n \geq m$. Then there exists $f \in S^*$ such that

$$||\sum_{k=n}^{r} A_k x_k|| = |f(\sum_{k=n}^{r} A_k x_k)|$$

$$= |\sum_{k=n}^{r} (A_k^* f)(x_k)|$$

$$\leq \sum_{k=n}^{r} ||A_k^* f||\ ||x_k||$$

$$\leq (\sum_{k=n}^{r} ||A_k^* f||^q)^{1/q} (\sum_{k=n}^{r} ||x_k||^p)^{1/p}$$

by Hölder's inequality. Hence

$$||\sum_{k=n}^{r} A_k x_k|| \leq H^{1/q} \varepsilon_n,$$

where $\varepsilon_n \to 0$ $(n \to \infty)$. Thus $\Sigma A_k x_k$ converges.

Conversely, $(A_k) \in \ell_p^\beta(X)$ implies $(A_k) \in \ell_1^\beta(X)$ so the existence of $m \in N$, such that $A_k \in B(X,Y)$ for $k \geq m$, follows from Proposition 3.7.

Now take any $f \in Y^*$. Then

$$(3.5) \qquad \sum_{k=m}^{\infty} (f, A_k x_k)$$

is a convergent series of complex numbers, for all $x \in \ell_p(X)$.

Also, for each $k \geq m$, there exists $z_k \in S$ such that

$$(3.6) \qquad ||A_k^* f|| \leq 2|(f, A_k z_k)|.$$

Let $\lambda = (\lambda_k) \in \ell_p$, so $||\lambda|| = (\Sigma|\lambda_k|^p)^{1/p}$. Then $(\lambda_k z_k) \in \ell_p(X)$, whence (3.5) implies that

$$\sum_{k=m}^{\infty} \lambda_k (f, A_k z_k)$$

converges for all $\lambda \in \ell_p$. Hence, by the classical result that $\ell_p^\beta = \ell_q$ it follows that, for all $f \in Y^*$,

$$\sum_{k=m}^{\infty} |(f, A_k z_k)|^q < \infty,$$

and so

$$\sum_{k=m}^{\infty} ||A_k^* f||^q < \infty,$$

by (3.6).

If we define, for $n \geq m$,

$$s_n(f) = \left(\sum_{k=m}^{n} ||A_k^* f||^q\right)^{1/q}$$

on the Banach space Y^* then each s_n is a continuous seminorm and so a version of the Banach-Steinhaus theorem (see e.g. Maddox [40], p.114) yields (3.4).

In the case $1 < p < \infty$ the α-dual of $\ell_p(X)$ is rather simpler than its β-dual, as we now show.

3.9 Proposition. Let $1 < p < \infty$. Then $(A_k) \in \ell_p^\alpha(X)$ if and only if there exists $m \in N$ such that $A_k \in B(X,Y)$ for all $k \geq m$ and

$$\sum_{k=m}^{\infty} ||A_k||^q < \infty$$

where $(1/p) + (1/q) = 1$.

Proof. The sufficiency is an immediate consequence of Hölder's inequality.

Conversely, $(A_k) \in \ell_p^\alpha(X)$ implies $(A_k) \in \ell_p^\beta(X)$ so the existence of the $m \in N$ in the proposition follows from Proposition 3.8.

Now for $k \geq m$ there exists z_k in S such that $2||A_k z_k|| \geq ||A_k||$.

For all $\lambda \in \ell_p$ we have $(\lambda_k z_k) \in \ell_p(X)$ and so

$$\sum_{k=m}^{\infty} |\lambda_k| \; ||A_k z_k|| < \infty$$

for all $\lambda \in \ell_p$. But $\ell_p^\alpha = \ell_q$ and so

$$H = \sum_{k=m}^{\infty} ||A_k z_k||^q < \infty,$$

whence

$$\sum_{k=m}^{\infty} ||A_k||^q \leq H.2^q,$$

which completes the proof.

We observe that, when $1 < p < \infty$, the inclusion

$$\ell_p^\alpha(X) \subset \ell_p^\beta(X)$$

may be strict. For example, take $X = Y = \ell_p$ and define $A_k \in B(X,Y)$ by

$$A_k x = (0,0,\dots,x_k,0,0,\dots)$$

with x_k in the k-position. Take $f \in S^*$. Then for $x \in \ell_p$,

$$f(x) = \Sigma f_i x_i$$

with $\Sigma|f_i|^q \le 1$, where $(1/p) + (1/q) = 1$. Hence

$$(A_k^* f)(x) = f_k x_k$$

and so $||A_k^* f|| = |f_k|$. Thus $\Sigma||A_k^* f||^q \le 1$, so (3.4) holds with $m = 1$, whence $(A_k) \in \ell_p^\beta(X)$. But $||A_k|| = 1$ for all k, so by Proposition 3.9 we see that (A_k) is not in $\ell_p^\alpha(X)$.

The following result relating $\ell_\infty^\beta(X)$ with unconditional convergence of ΣA_k is of interest in connection with the above ideas. The result was given by Thorp [74], though our approach is slightly different.

3.10 Proposition. *Let $A_k \in B(X,Y)$ for $k \in N$. Then $(A_k) \in \ell_\infty^\beta(X)$ implies ΣA_k is unconditionally convergent in the uniform operator topology. Also, if X is infinite dimensional then there exists ΣA_k unconditionally convergent in the uniform operator topology such that $(A_k) \notin \ell_\infty^\beta(X)$.*

Proof. Let $x \in X$, $\lambda \in \ell_\infty$ and $(A_k) \in \ell_\infty^\beta(X)$. Now

$$\left|\left|\sum_n^{n+p} \lambda_k A_k x\right|\right| \le ||R_n||.||\lambda||.||x||,$$

whence with the uniform operator norm,

$$\left|\left|\sum_n^{n+p} \lambda_k A_k\right|\right| \le ||R_n||.||\lambda||,$$

and since $||R_n|| \to 0$ $(n \to \infty)$ by Proposition 3.3, we see that ΣA_k is bounded multiplier convergent, and so is unconditionally convergent.

Suppose that X is infinite dimensional. Then X* is infinite dimensional, so the Dvoretzky-Rogers theorem yields a series Σf_k in X* which is unconditionally convergent but not absolutely convergent. Take $y \in Y$ with $||y|| = 1$ and define $A_k \in B(X,Y)$ for each $k \in N$ by

$$A_k x = f_k(x)y, \text{ for each } x \in X.$$

Then ΣA_k is unconditionally convergent, but $\Sigma||A_k|| = \Sigma||f_k||$ diverges. We cannot have $(A_k) \in \ell_\infty^\beta(X)$, for this would imply $\Sigma f_k(z_k)$ convergent for all $(z_k) \in \ell_\infty(X)$, whence $\Sigma||f_k|| < \infty$, by Proposition 3.5, which gives a contradiction.

We now turn to the determination of the operator β-dual of the space $w_p(X)$, where $0 < p < \infty$. Note that if $x_k \to \ell$ $(k \to \infty)$ then by the Cesàro mean theorem,

$$\frac{1}{n}\sum_{k=1}^{n}||x_k-\ell||^p \to 0 \ (n \to \infty),$$

and so $c(X) \subset w_p(X)$, whence $w_p^\beta(X) \subset c^\beta(X)$.

3.11 <u>Proposition</u>. <u>Let</u> $0 < p < 1$. <u>Then</u> $(A_k) \in w_p^\beta(X)$ <u>if and only if the</u> A_k <u>are, ultimately, bounded, and for some</u> $m \in N$,

$$\text{(3.7)} \qquad H = \sup \sum_{r=m}^{\infty} 2^{r/p} \max_r ||A_k^* f|| < \infty,$$

<u>where the supremum is taken over all</u> $f \in S^*$ <u>and</u> $\max_r$ <u>is over</u> $2^r \le k < 2^{r+1}$.

<u>Proof</u>. Suppose (3.7) holds and write $R_n = (A_n, A_{n+1}, \ldots)$. Take $i \ge m$ and $n \ge 2^i$. Then, where the supremum is over all $t \ge 2^i$ and all $x_k \in S$,

$$||R_n|| \le ||R_{2^i}|| = \sup||\sum_{k=2^i}^{t} A_k x_k||$$

$$\le \sup \sum_{k=2^i}^{t} |(f, A_k x_k)|, \text{ some } f \in S^*,$$

$$\leq \sup \sum_{k=2^i}^{t} ||A_k^* f||\ ||x_k||$$

$$\leq \sum_{r=i}^{\infty} \Sigma_r ||A_k^* f||$$

$$\leq H.2^{i/q}, \text{ where } q = p/(p-1) < 0.$$

In the above Σ_r denotes a sum over $2^r \leq k < 2^{r+1}$. It follows from the last inequality above that $||R_n|| \to 0$ $(n \to \infty)$, and so ΣA_k converges.

Now if

$$n^{-1} \sum_{k=1}^{n} ||x_k - \ell||^p \to 0 \ (n \to \infty)$$

then it is clear that

$$s_r = 2^{-r}\ \Sigma_r ||x_k - \ell||^p \to 0 \ (r \to \infty) \tag{3.8}$$

where the sum in (3.8) is over $2^k \leq k < 2^{r+1}$. Since $\Sigma A_k \ell$ converges we have only to show that $\Sigma A_k(x_k - \ell)$ converges. By an argument similar to that above, which showed that $||R_n|| \to 0$, we have

$$||\sum_{k=n}^{t} A_k(x_k-\ell)|| \leq \sum_{r=i}^{\infty} \Sigma_r ||A_k^* f||\ ||x_k - \ell||$$

$$\leq \sum_{r=i}^{\infty} \max_r ||A_k^* f|| \Sigma_r ||x_k - \ell||$$

$$\leq \sum_{r=i}^{\infty} 2^{r/p} \max_r ||A_k^* f|| s_r^{1/p}.$$

It is now immediate from (3.7) and (3.8) that $\Sigma A_k(x_k - \ell)$ converges.

For the necessity, if $(A_k) \in w_p^{\beta}(X)$ then $(A_k) \in c^{\beta}(X)$ so by Proposition 3.2 the A_k are, ultimately, bounded. Hence there exists m such that $A_k \in B(X,Y)$ for all $k \geq 2^m$. Take $f \in Y^*$. Then

$$\sum_{k=2^m}^{\infty} (f, A_k x_k)$$

converges for all $x \in w_p(X)$. Choose $z_k \in S$ such that $||A_k^* f|| \leq 2|(f, A_k z_k)|$ and take any complex sequence (a_k) such that $a_k \to 0\ (w_p)$. Then $(a_k z_k) \in w_p(X)$ and so

$$\sum_{k=2^m}^{\infty} a_k (f, A_k z_k)$$

converges whenever $a_k \to 0(w_p)$. It follows from Maddox [39] that

$$\sum_{r=m}^{\infty} 2^{r/p} \max_r |(f, A_k z_k)| < \infty$$

whence

$$\sum_{r=m}^{\infty} 2^{r/p} \max_r ||A_k^* f|| < \infty$$

for each $f \in Y^*$. Finally, arguing as in the necessity of Proposition 3.8 we obtain (3.7), and the proof is complete.

Using now familiar arguments there is little difficulty in proving:

3.12 Proposition. Let $0 < p < 1$. Then $(A_k) \in w_p^{\alpha}(X)$ if and only if the A_k are, ultimately, bounded, and for some $m \in N$,

$$\sum_{r=m}^{\infty} 2^{r/p} \max_r ||A_k|| < \infty.$$

3.13 Proposition. Let $1 \leq p < \infty$. Then $(A_k) \in w_p^{\beta}(X)$ if and only if the A_k are, ultimately, bounded, and for some $m \in N$,

$$\sup \sum_{r=0}^{\infty} 2^{r/p} (\Sigma_r ||A_k^* f||^q)^{1/q} < \infty, \tag{3.9}$$

and ΣA_k converges. In (3.9) the supremum is over all $f \in S^*$, the inner sum is over $2^r \le k < 2^{r+1}$, and $(1/p) + (1/q) = 1$.

The α-dual of $w_p(X)$ is also readily obtained when $1 \le p < \infty$.

With the restriction that all $A_k \in B(X,Y)$, the Propositions 3.11, 3.13 were given by Maddox [41].

In Proposition 3.13 we note that (3.9) is independent of ΣA_k convergent. For example, in the space of complex numbers, if $a_k = (-1)^k/k$ then Σa_k converges, but a simple calculation gives

$$2^{r/p}(\Sigma_r |a_k|^q)^{1/q} > 2^{-1}$$

so that (3.9) fails.

On the other hand, let

$$X = Y = \{x \in w_p : x_k \to 0 \ (w_p)\}.$$

Then X becomes a Banach space with

$$||x|| = \sup \ (2^{-r}\Sigma_r |x_k|^p)^{1/p}$$

where the supremum is over $r \ge 0$ and the sum is over $2^r \le k < 2^{r+1}$.

Now define $A_k \in B(X,Y)$ by

$$A_k x = (0,0,\ldots,x_1,0,0,\ldots), \text{ with } x_1 \text{ in the k-position.}$$

Take $f \in S^*$. Then for $x \in w_p$,

$$f(x) = \Sigma a_k x_k,$$

$$||f|| = \sum_{r=0}^{\infty} 2^{r/p}(\Sigma_r |a_k|^q)^{1/q} < \infty.$$

The representation for f, and the value of $||f||$ are to be found in Maddox [41]. See also Borwein [6].

Now $||A_k^*f|| = |a_k|$, and so (3.9) holds. But if $x = (1,0,0,\ldots)$ and we write

$$y(n) = \sum_{k=1}^{n} A_k x$$

then $||y(2^{r+1}-1) - y(2^r-1)|| = 1$, which implies that $\Sigma A_k x$ diverges.

4. Characterization of matrix classes

Operator versions of the major classical theorems on matrix transformations are proved. For example, Theorem 4.2 is the generalization of the Kojima-Schur theorem, and Corollary 4.3 generalizes the theorem of Toeplitz. Also more recent results on the characterization of some important matrix classes will be given.

We use the notation (E,F) for a matrix class, as in Definition 2. . For example, if A_{nk} are linear, but not necessarily bounded operators on a Banach space X into a Banach space Y then the assertion that the infinite matrix

$$A = (A_{nk}) \in (c(X),\ \ell_\infty(Y))$$

means that

$$A_n(x) = \Sigma A_{nk} x_k$$

converges in the norm of Y for every n and every $x \in c(X)$, and that $\sup_n ||A_n(x)|| < \infty$ for every $x \in c(X)$. In the sequel we shall use the following interesting result on unbounded operators which is due to Lorentz and Macphail [37].

4.1 Theorem. Let E and F be Banach spaces and (M_n) a decreasing sequence of closed linear subspaces of E. For each n let T_n be a linear operator on E into F such that T_n is bounded on M_n.

Suppose also that $(T_n x)$ is bounded for each $x \in E$. Then there exists $p \in N$ such that all the T_n are bounded on M_p.

Proof. Suppose there is no such p. Then for every M_p there exists T_n which is not bounded on M_p, where necessarily $n > p$. By passing to a subsequence we may assume without loss of generality that T_{n+1} is not bounded on M_n.

We shall obtain a contradiction by constructing an $x \in E$ such that $||T_{n+1}x|| > n$ for all $n \geq 2$.

Let us write

$$H_n = \sup\{||T_n x|| : x \in M_n \text{ and } ||x|| \leq 1\},$$

and define

$$\alpha_i = 1/2^i(1 + H_1 + H_2 + \ldots + H_i).$$

Then

$$(4.1) \qquad H_{n+1} \sum_{i=n+1}^{\infty} \alpha_i < 1.$$

Choose $x_1 \in M_1$ with $||x_1|| \leq 1$. Now T_3 is not bounded on M_2 so there exists $x_2 \in M_2$ with $||x_2|| \leq 1$ such that

$$\alpha_2||T_3x_2|| > 3 + \alpha_1||T_3x_1||.$$

Proceeding inductively we determine $x_n \in M_n$ with $||x_n|| \leq 1$ such that for $n \geq 2$,

$$(4.2) \qquad \alpha_n||T_{n+1}x_n|| > n + 1 + \sum_{i=1}^{n-1} \alpha_i||T_{n+1}x_i||.$$

Define x to be the sum of the absolutely convergent series $\Sigma\alpha_i x_i$. Then $x \in E$ and

$$||x|| \leq \Sigma\alpha_i \leq 1.$$

Also, $T_{n+1}x$ is equal to

$$\alpha_n T_{n+1}x_n + \sum_{i=1}^{n-1} \alpha_i T_{n+1}x_i + T_{n+1} \sum_{i=n+1}^{\infty} \alpha_i x_i.$$

Hence, by (4.2),

$$||T_{n+1}x|| > n + 1 - ||T_{n+1} \sum_{i=n+1}^{\infty} \alpha_i x_i||.$$

But $x_i \in M_{n+1}$ for $i \geq n + 1$, and since M_{n+1} is a closed linear subspace we have

$$\sum_{i=n+1}^{\infty} \alpha_i x_i \in M_{n+1}.$$

Hence, since T_{n+1} is bounded on M_{n+1} we see by (4.1) that

$$||T_{n+1} \sum_{i=n+1}^{\infty} \alpha_i x_i|| < 1.$$

Thus $||T_{n+1}x|| > n$, and the proof is complete.

We now prove a result, essentially due to Robinson [66], but see also Melvin-Melvin [55], which initiated the study of infinite matrices of linear operators in Banach spaces. Our proof follows the method of Lorentz and Macphail [37]; see also Zeller [84] for the case in which all the operators are bounded.

4.2 Theorem. Let A_{nk} be linear, but not necessarily bounded, operators on a Banach space X into a Banach space Y. Then

$$A = (A_{nk}) \in (c(X), c(Y))$$

if and only if there exists $m \in N$ such that

(4.3) $\sup_n ||(A_{nm}, A_{n,m+1}, \ldots)|| < \infty$,

(4.4) there exists $\lim_n A_{nk}$ for each k,

(4.5) ΣA_{nk} converges for each n,

(4.6) there exists $\lim_n \Sigma A_{nk}$.

Proof. Necessity. The necessity of (4.4) follows by considering the sequence which has x in the k-place and Θ elsewhere, and the necessity of (4.5) and (4.6) follows by considering the sequence (x, x, x, ...).

Now for the necessity of (4.3) : by Proposition 3.2, for each $n \in N$ there exists $f(n) \in N$ such that $A_{nk} \in B(X,Y)$ for all $k \geq f(n)$. Hence we may determine a strictly increasing sequence $(g(n))$ of natural numbers such that

$$(4.7) \qquad A_{nk} \in B(X,Y) \text{ for all } k \geq g(n).$$

Define, for each n,

$$L_n = \{x \in c(X) : x_k = \Theta \text{ for } k \leq n\},$$

so that L_n is a closed linear subspace of $c(X)$. If we write

$$M_n = L_{g(n)}$$

than (M_n) is decreasing.

Now $c(X)$ is a Banach space with $||x|| = \sup||x_k||$ for $x = (x_k) \in c(X)$, and each T_n given by

$$T_n x = \Sigma A_{nk} x_k$$

defines a linear operator on $c(X)$ into Y. If $x \in M_n$ then by (4.7) and the Banach-Steinhaus theorem we see that T_n is bounded on M_n. Thus the hypotheses of Theorem 4.1 are satisfied with $E = c(X)$ and $F = Y$, so that there exists $p \in N$ such that all the T_n are bounded on M_p. Again by the Banach-Steinhaus theorem, there is a constant H such that

$$||T_n x|| \leq H||x||$$

for all n and all $x \in M_p$.

By suitable choice of x, essentially as in Proposition 3.1, we now see that (4.3) holds with $m = 1 + g(p)$.

<u>Sufficiency</u>. Let $x_k \to \ell (k \to \infty)$ and denote the supremum in (4.3) by H. Given $\varepsilon > 0$ we have $||x_k - \ell|| < \varepsilon$ for all sufficiently large k, and so for all sufficiently large r, all $s \geq r$, and all $n \in N$,

$$(4.8) \qquad ||\sum_{k=r}^{s} A_{nk}(x_k - \ell)|| \leq H\varepsilon,$$

whence $\Sigma A_{nk}(x_k - \ell)$ converges for all $n \in N$.

Applying (4.4) to (4.8) we see that $\Sigma \lim_n A_{nk}(x_k - \ell)$ converges. If we now write $A_k = \lim_n A_{nk}$ for each k, and choose $r > m$ such that $||x_k - \ell|| < \varepsilon$ for all $k \geq r$ then

$$||\Sigma(A_{nk} - A_k)(x_k - \ell)||$$

$$\leq \sum_{k<r} ||(A_{nk} - A_k)(x_k - \ell)|| + ||\sum_{k \geq r} (A_{nk} - A_k)(x_k - \ell)||$$

$$\leq \sum_{k<r} ||(A_{nk} - A_k)(x_k - \ell)|| + 2H\varepsilon.$$

It follows from (4.4) that

$$\Sigma(A_{nk} - A_k)(x_k - \ell) \to \Theta \quad (n \to \infty)$$

and consequently

$$\lim_n \Sigma A_{nk} x_k = \lim_n \Sigma A_{nk} \ell + \Sigma \lim_n A_{nk}(x_k - \ell),$$

which completes the proof.

As a corollary we immediately obtain the operator version of the classical theorem of Toeplitz which characterizes complex matrices which map convergent sequences into convergent sequences with the same limit.

4.3 <u>Corollary</u>. <u>Let</u> A_{nk} <u>be linear operators on a Banach space</u> X <u>into itself</u>. <u>Then</u>

$$\lim_n \Sigma A_{nk}x_k = \lim x_k$$

for all $x = (x_k) \in c(X)$ *if and only if there exists* $m \in N$ *such that*

(4.9) $$\sup_n ||(A_{nm}, A_{n,m+1}, \ldots)|| < \infty,$$

(4.10) $$\lim_n A_{nk} = 0 \text{ for each } k,$$

(4.11) $$\Sigma A_{nk} \text{ converges for each } n,$$

(4.12) $$\lim_n \Sigma A_{nk} = I,$$

where I *is the identity operator.*

We remark that Ramanujan [63] has given an extension of Theorem 4.2 where Banach spaces are replaced by Fréchet spaces, but with the A_{nk} restricted to be continuous linear operators between the Fréchet spaces.

Our next result is an operator version of the classical theorem of Steinhaus - see Cooke [9].

4.4 *Theorem.* *Let* $A_{nk} \in B(X,X)$, *where* X *is a Banach space, be such that* $A = (A_{nk})$ *is a Toeplitz matrix, so the conditions of Corollary 4.3 hold. Then there is a bounded sequence which is not summable* A.

Proof. In view of the note after Proposition 3.3 it may be the case that there exist n and $x \in \ell_\infty(X)$ such that $\Sigma A_{nk}x_k$ diverges, whence the conclusion of our theorem is trivial. In the contrary case we have $(A_{nk})_{k\in N} \in \ell_\infty^\beta(X)$ for each n, so by Proposition 3.3,

(4.13) $$\lim_k ||(A_{nk}, A_{n,k+1}, \ldots)|| = 0.$$

Now fix $z \in U$. Then $||\Sigma A_{nk}z|| \to 1$ $(n \to \infty)$ so there exists n(1) such that

$$||\Sigma A_{nk}z|| > 3/4 \text{ for all } n \geq n(1).$$

By (4.13) there exists k(1) such that

$$||(A_{n(1),1+k(1)}, A_{n(1),2+k(1)}, \ldots)|| < 1/12$$

and so

$$||\sum_{1+k(1)}^{\infty} A_{n(1)k}y_k|| \leq ||y||/12$$

for each $y \in \ell_\infty(X)$.

Define $x_k = z$ for $1 \leq k \leq k(1)$. Then there exists $n(2) > n(1)$ such that

$$\sum_{k=1}^{k(1)} ||A_{n(2)k}z|| < 1/6$$

and $k(2) > k(1)$ such that

$$||(A_{n(2),1+k(2)}, A_{n(2),2+k(2)}, \ldots)|| < 1/6.$$

Define $x_k = \Theta$ for $k(1) < k \leq k(2)$. Proceeding inductively we construct $k(1) < k(2) < k(3) < \ldots$, $n(1) < n(2) < n(3) < \ldots$, and a bounded sequence $x = (x_k)$ which takes only the values z and Θ. The construction yields a sequence

$$(A_n(x)) = (\Sigma A_{nk}x_k)$$

which is not Cauchy. For example,

$$||A_{n(1)}(x) - A_{n(2)}(x)||$$

is equal to

$$||\Sigma A_{n(1)k}z - \sum_{1+k(1)}^{\infty} A_{n(1)k}(z-x_k) - \sum_{1}^{k(1)} A_{n(2)k}z - \sum_{1+k(2)}^{\infty} A_{n(2)k}x_k||$$

$$> 3/4 - 1/12 - 1/6 - 1/6 = 1/3.$$

Using Theorem 4.2, we may characterize those series to sequence transformations which leave invariant the sum of each convergent series.

4.5 Theorem. *Let* B_{nk} *be linear operators on Banach space X into itself. Then*

$$\lim_n \Sigma B_{nk} a_k = \Sigma a_k$$

for all convergent series Σa_k, *if and only if there exists* $m \in N$ *such that*

(4.14) $$M = \sup_n ||B_{nm}|| < \infty$$

(4.15) $$H = \sup_n ||(\Delta B_{nm}, \Delta B_{n,m+1}, \ldots)|| < \infty,$$

(4.16) $$\lim_n B_{nk} = I \text{ for each } k.$$

In (4.15) *we define* $\Delta B_{nk} = B_{nk} - B_{n,k+1}$.

Proof. Sufficiency. Let $\Sigma a_k = s$ and define $c_1 = a_1 - s$, $c_k = a_k$ otherwise. Then

$$\sum_{k=1}^{q} B_{nk} a_k = B_{n1} s + \sum_{k=1}^{q} B_{nk} c_k$$

and once we have shown $\Sigma B_{nk} c_k$ converges for each n and has sum $\Sigma \Delta B_{nk} s_k$, where

$$s_k = c_1 + c_2 + \ldots + c_k,$$

it follows from (4.15) and (4.16) that $\lim_n \Sigma \Delta B_{nk} s_k = \Theta$, whence

$$\lim_n \Sigma B_{nk} a_k = \lim_n B_{n1} s = s.$$

Now by Abel's partial summation

(4.17) $$\sum_{r}^{r+p} B_{nk} c_k = B_{n,r+p} s_{r+p} - B_{nr} s_{r-1} + \sum_{r}^{r+p-1} \Delta B_{nk} s_k.$$

Let $\varepsilon > 0$. Then for all sufficiently large r it follows from (4.15) that

$$(4.18) \qquad \left\| \sum_{r}^{r+p-1} \Delta B_{nk} s_k \right\| \le H\varepsilon.$$

Also, we have, for any z,

$$B_{n,m+p} z = B_{nm} z - \sum_{m}^{m+p-1} \Delta B_{nk} z$$

and so for all n and all $p \ge 0$, (4.14), (4.15) imply $||B_{n,m+p}|| \le M + H$, so for all sufficiently large r, we see from (4.17), (4.18) that $\Sigma B_{nk} c_k$ converges. But (4.17) is valid for $r = 1$, on setting $s_0 = \theta$, so letting $p \to \infty$ we have $\Sigma B_{nk} c_k = \Sigma \Delta B_{nk} s_k$.

Necessity. An obvious choice of (a_k) yields the necessity of (4.16). Now write

$$s_k = a_1 + a_2 + \dots + a_k.$$

If (B_k) is a sequence of linear operators on X into itself and $\Sigma B_k a_k$ converges for all convergent Σa_k then, by Abel's partial summation and an application of Theorem 4.2, there exists $p \in N$ such that

$$||B_p|| < \infty \text{ and } ||(\Delta B_p, \Delta B_{p+1}, \dots)|| < \infty.$$

These last two conditions imply

$$||B_k|| \le ||B_p|| + ||(\Delta B_p, \dots)||$$

for all $k > p$. Hence

$$\Sigma B_k a_k = \Sigma \Delta B_k s_k$$

for all Σa_k convergent to θ.

Thus for each n we have

$$\Sigma B_{nk} a_k = \Sigma \Delta B_{nk} s_k$$

and $(\Delta B_{nk}) \in (c_0(X), c_0(X))$, whence by the argument of the necessity of Theorem 4.2 there exists $q \in N$ such that

$$(4.19) \qquad \sup_n || (\Delta B_{nq}, \Delta B_{n,q+1}, \ldots) || < \infty.$$

Now $\Sigma B_{nk} a_k$ converges for each n and for all Σa_k which converge, and so $\Sigma B_{nk} a_k$ converges on $\ell_1(X)$ so by Proposition 3.7 there exists a strictly increasing sequence $(g(n))$ of natural numbers such that $B_{nk} \in B(X,X)$ for all $k \geq g(n)$. Then each T_n given by

$$T_n a = \Sigma B_{nk} a_k$$

defines a linear operator on

$$\gamma(X) = \{a : \Sigma a_k \text{ converges}\}$$

into X. Note that $\gamma(X)$ is a Banach space under the norm

$$||a|| = \sup_k ||s_k||.$$

Let $L_n = \{a \in \gamma(X) : a_k = \Theta \text{ for } k \leq n\}$, so that L_n is a closed linear subspace of $\gamma(X)$. Writing $M_n = L_{g(n)}$, and noting that T_n is bounded on M_n we may invoke Theorem 4.1 and the Banach-Steinhaus theorem to obtain

$$\sup_n ||T_n a|| \leq D ||a||, \text{ on } M_p,$$

for some $p \in N$ and some constant D.

Define $m = 1 + g(p) + q$. Then (4.19) implies (4.15).

Now take $x \in U$ and define $a_m = x$, $a_k = \Theta$ otherwise. Then

$$\sup_n ||B_{nm}|| \leq D$$

which implies (4.14). This completes the proof.

The next theorem, proved by Maddox [43], extends the classical theorem of Schur on matrices which map bounded sequences into convergent sequences. For simplicity we restrict our attention to infinite matrices whose elements are bounded linear operators.

4.6 Theorem. *Let X and Y be Banach spaces and* $A_{nk} \in B(X,Y)$. *Write for each* $n,m \in N$,

$$R_{nm} = (A_{nm}, A_{n,m+1}, \ldots)$$

so that R_{nm} *is the* m-th *tail of the* n-th *row of the matrix* $A = (A_{nk})$.

Then $A \in (\ell_\infty(X), c(Y))$ *if and only if*

(4.20) *there exists* $\lim_n A_{nk} = A_k$, *for each* k,

(4.21) $\lim_m \|R_{nm}\| = 0$, *for each* n,

(4.22) $\sup_n \|R_{nm} - R_m\| \to 0 \ (m \to \infty)$,

where $R_m = (A_m, A_{m+1}, \ldots)$.

When (4.20) - (4.22) *hold we have*

(4.23) $\lim_n \Sigma A_{nk} x_k = \Sigma A_k x_k$

for each $x = (x_k) \in \ell_\infty(X)$.

It is to be understood in (4.21) that $\|R_{n1}\| < \infty$ *for each* n.

Proof. *Sufficiency*. By (4.20) and the Banach-Steinhaus theorem, each $A_{nk} \in B(X,Y)$. Now if $n > m$ and $x \in \ell_\infty(X)$ with $\|x\| = \sup\|x_k\| < \infty$, then

$$\left\|\sum_m^n A_k x_k\right\| \le \left\|\sum_m^n (A_k - A_{1k}) x_k\right\| + \left\|\sum_m^n A_{1k} x_k\right\|$$

$$\leq ||x|| \, ||R_m - R_{1m}|| + ||x|| \, ||R_{1m}||$$

$$\to 0 \ (m \to \infty)$$

by (4.21), (4.22). Hence $\Sigma A_k x_k$ converges.

By (4.21), $\Sigma A_{nk} x_k$ converges for each n and each $x \in \ell_\infty(X)$.

Now let $\varepsilon > 0$ and choose $m = m(\varepsilon) > 1$ by (4.22) such that $\sup_n ||R_{nm} - R_m|| < \varepsilon$. Then for each $x \in \ell_\infty(X)$, and each n,

$$||\Sigma A_{nk} x_k - \Sigma A_k x_k||$$

$$\leq \sum_{k<m} ||(A_{nk} - A_k) x_k|| + \varepsilon ||x||.$$

Letting $n \to \infty$ we see by (4.20) that (4.23) holds.

<u>Necessity</u>. It is trivial that (4.20) is necessary. Now for each n we have $R_{n1} \in \ell_\infty^\beta(X)$. Hence by Proposition 3.3, since we are assuming that all the A_{nk} are bounded, we have $||R_{n1}|| < \infty$ for each n and (4.21).

Let us denote by V the closed unit sphere in $\ell_\infty(X)$ and define

$$d(x,y) = \Sigma ||x_k - y_k|| . 2^{-k}$$

for each x and y in V. Then (V,d) is a complete metric space, for d is obviously a metric on V, and if $(x^{(n)})$ is a Cauchy sequence in (V,d) then for each k,

$$||x_k^{(n)} - x_k^{(m)}|| \leq 2^k d(x^{(n)}, x^{(m)})$$

and so

$$||x_k^{(n)} - x_k^{(m)}|| \to 0 \ (m,n \to \infty).$$

Since X is a Banach space, there exists, for each k,

$$\lim_m x_k^{(m)} = x_k, \text{ say,}$$

and since $||x_k^{(m)}|| \leq 1$ for all m,k, we have $||x_k|| \leq 1$, so $x = (x_k) \in V$.

Moreover, for each p,

$$\sum_{k=1}^{p} ||x_k^{(n)} - x_k^{(m)}||2^{-k} \leq d(x^{(n)},x^{(m)}),$$

and so

$$\sum_{k=1}^{\infty} ||x_k^{(n)} - x_k||2^{-k} \leq \limsup_m d(x^{(n)},x^{(m)})$$

for each n. Applying the operator $\limsup_n$ we deduce that $(x^{(n)})$ converges to x.

Now for each $x \in V$ and each (m,n) define

$$f_{mn}(x) = \Sigma(A_{mk} - A_{nk})x_k.$$

Then for each p > 1, and x, y in V,

$$||f_{mn}(x-y)|| \leq d(x,y).2^p \sum_{k<p} ||A_{mk} - A_{nk}||$$

$$+ 2(||R_{mp}|| + ||R_{np}||).$$

It follows from (4.21) that f_{mn} is continuous on V.

Let $\varepsilon > 0$ and write

$$F_{mn} = \{x \in V : ||f_{mn}(x)|| \leq \varepsilon\}.$$

The continuity of f_{mn} implies that each F_{mn} is closed, so for each $p \in N$,

$$E_p = \cap F_{mn}$$

is closed, where the intersection is over all $m \geq p$ and all $n \geq p$.

Since $(\Sigma A_{nk}x_k) \in c(Y)$ for all $x \in V$ it follows readily that

$$V = \cup\{E_n : n \in N\}.$$

By the Baire category theorem, see for example Maddox [40], there exists $p \in N$ and a sphere $S(a,r)$ in (V,d) such that $S(a,r) \subset E_p$. Now choose $i > 1$ such that

$$\sum_{k\geq i} 2^{-k} < r/2.$$

Take any $x \in V$ and define $y_k = a_k$ for $k < i$, $y_k = x_k$ for $i \leq k \leq i + j$, and $y_k = \theta$ otherwise. Then

$$d(y,a) = \sum_{i}^{i+j} ||x_k-a_k||2^{-k} + \sum_{i+j+1}^{\infty} ||a_k||2^{-k}$$

$$\leq 2 \sum_{k\geq i} 2^{-k} < r,$$

whence $y \in E_p$, and so for all $m,n \geq p$,

$$||\sum_{k<i} (A_{mk}-A_{nk})a_k + \sum_{k=i}^{i+j} (A_{mk}-A_{nk})x_k|| \leq \varepsilon.$$

Hence, by (4.20) there exists M such that

$$||\sum_{k=i}^{i+j} (A_{mk}-A_{nk})x_k|| < 2\varepsilon$$

for all $m,n > M$. Letting $m \to \infty$, and then taking the supremum over j, we obtain, for all $n > M$, $||R_{ni}-R_i|| \leq 2\varepsilon$, whence $||R_{nj}-R_j|| \leq 2\varepsilon$ for all $n > M$ and all $j \geq i$.

Now consider n such that $1 \leq n \leq M + 1$. By (4.21) there exists q such that

$$||R_{nj}|| < \varepsilon, \text{ for all } j > q.$$

Let p be any natural number with $p > i + q$. Then with $n = M + 1$,

$$||R_p|| \le ||R_{np}-R_p|| + ||R_{np}||$$

$$\le 2\varepsilon + ||R_{np}|| < 3\varepsilon.$$

Hence, for $1 \le n \le m + 1$,

$$||R_{np}-R_p|| \le ||R_{np}|| + ||R_p|| < 4\varepsilon.$$

It follows that $||R_{np}-R_p|| < 4\varepsilon$ for all $n \in N$ and all $p > i + q$, whence (4.22) holds. This completes the proof.

If $X = Y = C$ and the A_{nk} are identified with complex numbers a_{nk} then

$$||R_{nm}|| = \sum_{k=m}^{\infty} |a_{nk}|.$$

It follows readily that the conditions of the theorem reduce to (2.7) and (2.13), so by the remark after Theorem 2.10 the conditions become those of the classical form of Schur's theorem.

We shall continue to study matrix maps of $\ell_\infty(X)$ by considering the seminorm p on $\ell_\infty(X)$:

$$p(x) = \lim\sup||x_k||.$$

If q is a functional on $\ell_\infty(X)$ and $M \ge 0$ is a real number we shall write

$$p \le Mq$$

to mean that $p(x) \le Mq(x)$ for all $x \in \ell_\infty(X)$. If $A_{nk} \in B(X,Y)$ and $A = (A_{nk})$ we write pA for p_0A. In Theorem 4.7 we give necessary and sufficient conditions for $pA \le Mp$ and use them to characterize the class $(\ell_\infty(X), c_0(Y))$. It is understood that the norm involved in p(Ax) is that of Y. Theorems 4.7,

4.8 appear in Maddox [45].

4.7 Theorem. Let X and Y be Banach spaces and $A_{nk} \in B(X,Y)$. Define R_{nm} as in Theorem 4.6. Then $pA \le Mp$ if and only if

(4.24) $$\lim_n A_{nk} = 0 \text{ for each } k,$$

(4.25) $$\|R_{n1}\| < \infty \text{ and } \lim_m \|R_{nm}\| = 0 \text{ for each } n,$$

(4.26) $$\lim_m \limsup_n \|R_{nm}\| \le M.$$

Proof. Sufficiency. Let (4.24) - (4.26) hold and take $x \in \ell_\infty(X)$. By (4.25), $y_n = \Sigma A_{nk}x_k$ exists for each n, and if $m > 1$ we have

$$\|y_n\| \le \sum_{k<m} \|A_{nk}x_k\| + \|R_{nm}\| . \sup_{k\ge m}\|x_k\|.$$

Applying the operator $\lim_m \limsup_n$ to both sides of this inequality, (4.24) and (4.26) imply that $p(Ax) \le p(x)$.

Necessity. Suppose $pA \le Mp$. Take any $z \in X$ and define $x_k = z$, $x_n = \theta$ $(n \ne k)$. Then $p(x) = 0$, which implies (4.24).

The convergence for each n of $\Sigma A_{nk}x_k$ on $\ell_\infty(X)$ implies, by Proposition 3.3, that (4.25) holds.

Since $\sup_n\|y_n\| < \infty$ on $\ell_\infty(X)$ the Banach-Steinhaus theorem implies that $\sup_n\|R_{n1}\| < \infty$. Also, we have $\|R_{n,m+1}\| \le \|R_{nm}\|$, so that the limit in (4.26) exists. Denote this limit by H. If $H = 0$ then (4.26) is obviously true. Suppose then that $H > 0$ and write

$$a(m) = \limsup_n \|R_{nm}\|.$$

Then $a(1) \ge H$, and there exists $n(1)$ such that $\|R_{n(1)1}\| > H - H/4$. By definition of the group norm $\|R_{n(1)1}\|$ there exist $k(1)$ and $z_1,\ldots,z_{k(1)} \in S$

such that

$$\left\|\sum_{k=1}^{k(1)} A_{n(1)k}z_k\right\| > \|R_{n(1)1}\| - H/4.$$

Thus

$$\left\|\sum_{k=1}^{k(1)} A_{n(1)k}z_k\right\| > H - H/2,$$

which implies that not all the z_k are Θ. Let

$$d = \max\{\|z_k\| : z_k \neq \Theta,\ 1 \le k \le k(1)\},$$

and define $x_k = z_k/d$ if $z_k \neq \Theta$, and $x_k = \Theta$ if $z_k = \Theta$. Then, since $0 < d \le 1$,

$$\left\|\sum_{k=1}^{k(1)} A_{n(1)k}x_k\right\| > H-H/2, \tag{4.27}$$

where all the $x_k \in S$, and at least one $x_k \in U$.

Now $a(k(1) + 1) \ge H$, and since

$$\lim_n \left\|\sum_{k=1}^{k(1)} A_{nk}x_k\right\| = 0,$$

there exists $n(2) > n(1)$ such that

$$\|R_{n(2),k(1)+1}\| > H - H/2^3,$$

$$\left\|\sum_{k=1}^{k(1)} A_{n(2)k}x_k\right\| < H/2^2.$$

By the argument leading to (4.27), there exist $q > k(1)$ and $x_{k(1)+1}, \ldots, x_q \in S$, with at least one $x_k \in U$ such that

$$\left\|\sum_{k(1)+1}^{q} A_{n(2)k}x_k\right\| > H - H/2^2.$$

Since (4.25) holds we may choose $k(2) > q$ such that

$$\|R_{n(2),k(2)+1}\| < H/2^2.$$

Define $x_k = \theta$ for $q < k \le k(2)$, so that

$$||\sum_{k(1)+1}^{k(2)} A_{n(2)k}x_k|| > H-H/2^2.$$

Proceeding inductively we obtain $x = (x_k)$ with $||x_k|| \le 1$ for all k and $||x_k|| = 1$ for infinitely many k. Hence we have $p(x) = 1$, $n(i) < n(i+1)$ and $k(i) < k(i+1)$. If $y_n = \Sigma A_{nk}x_k$ then for $i > 1$ and $n = n(i)$,

$$||y_n|| = ||\sum_{k(i-1)+1}^{k(i)} + \sum_{1}^{k(i-1)} + \sum_{k(i)+1}^{\infty}||$$

$$> H-H/2^i - H/2^i - ||R_{n,k(i)+1}||$$

$$> H - 3H/2^i.$$

Hence $\lim\sup_n ||y_n|| \ge H$ and so $Mp(x) \ge p(Ax) \ge H$, which yields $M \ge H$ since $p(x) = 1$. This completes the proof.

4.8 Theorem. $A \in (\ell_\infty(X), c_0(Y))$ if and only if the conditions of Theorem 4.7 hold with $M = 0$.

Proof. We merely take $M = 0$ in Theorem 4.7.

It is clear that, in the case $X = Y = C$, the conditions of Theorem 4.8 reduce to those of Theorem 2.11, with the usual identification regarding A_{nk}.

The case $p = 1$ of the next theorem was stated without proof by Lorentz and MacPhail [37]. Also, a generalization to Fréchet spaces of the result concerning $(\ell_1(X), \ell_1(Y))$ was given by Wood [81]. In Theorem 4.9 we suppose X and Y are Banach spaces and that $A_{nk} \in B(X,Y)$.

4.9 Theorem. Let $1 \le p < \infty$. Then $A \in (\ell_1(X), \ell_p(Y))$ if and only if

$$\sup \sum_{n=1}^{\infty} ||A_{nk}z||^p < \infty,$$

where the supremum is taken over all $z \in U$ *and all* $k \in N$,

Proof. *Necessity*. Define, on $\ell_1(X)$,

$$B_{nr}(x) = \sum_{k=1}^{r} A_{nk}x_k, \text{ where } x = (x_k) \in \ell_1(X).$$

Then for each n,r we have $B_{nr} \in B(\ell_1(X), Y)$. By hypothesis there exists on $\ell_1(X)$, for each n,

$$A_n(x) = \lim_r B_{nr}(x),$$

so the Banach Steinhaus theorem implies $A_n \in B(\ell_1(X), Y)$.

Now define

$$q_r(x) = \Big(\sum_{n=1}^{r} ||A_n(x)||^p\Big)^{1/p}.$$

Then (q_r) is a sequence of continuous seminorms on $\ell_1(X)$, so by a standard theorem (see e.g. Maddox [40], p.114) there exists a constant M such that

$$\sum_{n=1}^{\infty} ||A_n(x)||^p \leq M^p||x||^p, \tag{4.28}$$

on $\ell_1(X)$. Now take any $z \in U$ and any $k \in N$, and define $x = (\theta,\theta,\ldots,z,\theta,\theta,\ldots)$ with z in the k-position. Putting this x in (4.28) we obtain our result.

Sufficiency. If the condition in the theorem holds, then there is a positive constant M such that

$$\sum_{n=1}^{\infty} ||A_{nk}z||^p \leq M^p \tag{4.29}$$

for all $z \in U$, and all $k \in N$, which implies that the operator norms $||A_{nk}||$ do not exceed M for all n and k. Consequently, for each n and each $x \in \ell_1(X)$,

$$\sum_{k=1}^{\infty} A_{nk}x_k$$

is absolutely convergent.

Now let $x \in \ell_1(X)$. Then (4.29) implies

$$\sum_{n=1}^{\infty} ||A_{nk}x_k||^p \le M^p||x_k||^p \tag{4.30}$$

for each k. Hence for m, r $\in$ N, by Minkowski's inequality,

$$\left(\sum_{n=1}^{m} ||\sum_{k=1}^{r} A_{nk}x_k||^p\right)^{1/p} \le \sum_{k=1}^{r} \left(\sum_{n=1}^{m} ||A_{nk}x_k||^p\right)^{1/p}$$

$$\le \sum_{k=1}^{r} M||x_k||, \text{ by (4.30)},$$

$$\le M||x||.$$

Applying the operator $\lim_m \lim_r$ we see that

$$\sum_{n=1}^{\infty} ||\sum_{k=1}^{\infty} A_{nk}x_k||^p \le M^p||x||^p,$$

whence $A \in (\ell_1(X), \ell_p(Y))$, and the proof is complete.

Using similar reasoning we can deal with the case $p = \infty$ in Theorem 4.9:

4.10 Theorem. $A \in (\ell_1(X), \ell_\infty(Y))$ *if and only if*

$$\sup_{n,k}||A_{nk}|| < \infty.$$

4.11 Remark. In connection with Theorem 4.9 we note that the condition

$$\sup_k \sum_{n=1}^{\infty} ||A_{nk}||^p < \infty, \tag{4.31}$$

though sufficient for $A \in (\ell_1(X), \ell_p(Y))$ is not in general necessary. For example, consider the case $p = 1$ and let $X = Y = \ell_1$, the Banach space of absolutely convergent complex series. Define for each $z \in \ell_1$ and each $n,k \in N$,

$$A_{nk}z = (0,0,\ldots,z_n,0,0,\ldots),$$

with z_n in the n-position. It is clear that $A_{nk} \in B(\ell_1,\ell_1)$ and $||A_{nk}|| = 1$, so (4.31), with p = 1, is false. But if $z \in U$, i.e. $||z|| = \Sigma|z_k| = 1$, then for all k,

$$\sum_{n=1}^{\infty} ||A_{nk}z|| = \sum_{n=1}^{\infty} |z_n| = 1,$$

so that $A \in (\ell_1(\ell_1),\ \ell_1(\ell_1))$ by Theorem 4.9.

Theorem 4.16 below, due to Maddox and Wickstead [46], is an operator version of Crone's result on (ℓ_2,ℓ_2). The theorem extends Crone's theorem to infinite matrices (A_{ij}) of bounded linear operators on the Hilbert direct sum of a sequence of Hilbert spaces.

First we describe the basic ideas to be used. For each $i \in N$ let H_i be a Hilbert space with inner product $\langle .,. \rangle$. It is implicit that the inner product depends on i. By H we denote the space of all sequences $x = (x_i)$ with $x_i \in H_i$ and by $\ell_0(H)$ the subspace of all finite sequences.

Let D be the Hilbert direct sum of (H_i), that is

$$D = \bigoplus{}_2 H_i = \{x \in H : \Sigma_i ||x_i||^2 < \infty\}.$$

With the inner product

$$\langle x,y \rangle = \Sigma_i \langle x_i,y_i \rangle$$

we see that D is a Hilbert space, and for each $x \in D$,

$$||x|| = (\Sigma_i ||x_i||^2)^{1/2}.$$

In the special case when $H_i = H_1$ for all $i \in N$ we have $D = \ell_2(H_1)$.

If $T \in B(H_j,H_i)$ then by the Riesz-Fréchet theorem, there is a unique $T^* \in B(H_i,H_j)$ such that

$$\langle Tx_j,\ x_i \rangle = \langle x_j, T^*x_i \rangle$$

for all $x_i \in H_i$ and all $x_j \in H_j$.

We denote by M(H) the set of all matrices $A = (A_{ij})$ with $A_{ij} \in B(H_j,H_i)$. If A, B $\in$ M(H) we say that their product AB exists if all the series $\Sigma_k A_{ik}B_{kj}$ converge in the strong operator topology, i.e. $\Sigma_k A_{ik}B_{kj}y$ converges in the H_i norm for all $y \in H_j$. Then the (i,j) entry of AB is defined to be $\Sigma_k A_{ik}B_{kj}$. It follows from the Banach-Steinhaus theorem that if AB exists then AB $\in$ M(H).

We write P_n for the matrix whose (i,i) entry is the identity of H_i, for i = 1, 2, ...,n, and whose other entries are zero. Thus

$$P_n x = (x_1,x_2,\ldots,x_n,\theta,\theta,\theta,\ldots)$$

for each x $\in$ H.

By analogy with our earlier notation for matrix classes we denote by (D,D) the set of all A $\in$ M(H) which satisfy the two conditions:

(4.32) $\Sigma_j A_{ij}x_j$ converges in the H_i norm for all i and all x $\in$ D,

(4.33) $(\Sigma_j A_{ij}x_j)_{i\in N} \in D$ for all x $\in$ D.

Note that $P_n \in$ (D,D) for each n $\in$ N.

If A $\in$ (D,D) then, by the Banach-Steinhaus theorem, we see that (4.33) defines an element of B(D,D), and we also denote this element by A.

Conversely, one may show that every element of B(D,D) determines a matrix in (D,D).

Since B(D,D) is an algebra so is (D,D).

If A $\in$ M(H) then we define

(4.34) $$A^* = (A^*_{ji}).$$

Now if A $\in$ (D,D) and A is also the corresponding operator in B(D,D) then there is a unique operator A* $\in$ B(D,D) such that $\langle Ax,y\rangle = \langle x,A^*y\rangle$ for all x,y $\in$ D.

But A^* determines a matrix in (D,D), and we readily see that this matrix is given by (4.34).

Hence $A \in (D,D)$ implies $A^* \in (D,D)$.

A matrix $A \in M(H)$ is called Hermitian if and only if $A = A^*$ where A^* is defined by (4.34).

By $HM(H)$ we denote the set of all Hermitian matrices.

In order to prove the main results we give some lemmas.

4.12 Lemma. Let $A \in M(H)$. Then $A \in (D,D)$ if and only if A^*A exists and is in (D,D).

Proof. Sufficiency. Let $x \in \ell_0(H)$. Choose $n \in N$ such that $P_n x = x$. Now for each $m \in N$,

$$||P_m Ax||^2 = \langle P_m Ax, P_m Ax\rangle = \sum_{i=1}^{m} \langle \sum_{j=1}^{n} A_{ij}x_j, \sum_{k=1}^{n} A_{ik}x_k\rangle$$

$$= \sum_{j=1}^{n} \sum_{k=1}^{n} \langle x_j, \sum_{i=1}^{m} (A^*)_{ji}A_{ik}x_k\rangle.$$

Letting $m \to \infty$ we see that

$$||Ax||^2 = \sum_{j=1}^{n} \sum_{k=1}^{n} \langle x_j, \sum_{i=1}^{\infty} (A^*)_{ji}A_{ik}x_k\rangle$$

$$= \sum_{j=1}^{n} \langle x_j, \sum_{k=1}^{n} (A^*A)_{jk}x_k\rangle$$

$$= \langle x, A^*Ax\rangle.$$

It follows that A defines a bounded linear operator on $\ell_0(H)$, regarded as a normed subspace of D, into D. Also, the norm of A does not exceed $||A^*A||^{1/2}$.

Now if $x \in D$ then $\lim_n P_n x = x$ in the norm of D, whence the sequence $(AP_n x)$ is Cauchy in D, so that there exists $\lim_n AP_n x \equiv y \in D$. By coordinatewise convergence we must have, for each i,

$$y_i = \lim_n (AP_n x)_i = \lim_n \sum_{j=1}^{n} A_{ij} x_j = \Sigma_j A_{ij} x_j ,$$

whence (4.32) holds. Also, $y = (y_i) \in D$ and so (4.33) is valid, which means that $A \in (D,D)$.

<u>Necessity</u>. If $A \in (D,D)$ then $A^* \in (D,D)$, and since (D,D) is an algebra there exists A^*A and $A^*A \in (D,D)$. This proves the lemma.

4.13 <u>Lemma</u>. <u>If</u> $A \in M(H)$ <u>and</u> $n \in N$ <u>then</u>

$$||P_n AP_n x|| \le n.\ \max\{||A_{ij}|| : i,j \le n\}.$$

<u>Proof</u>. If $x \in D$, then the Cauchy-Schwarz inequality yields

$$||P_n AP_n x||^2 = \sum_{i=1}^{n} || \sum_{j=1}^{n} A_{ij} x_j ||^2$$

$$\le \sum_{i=1}^{n} (\sum_{j=1}^{n} ||A_{ij}||^2) (\sum_{j=1}^{\infty} ||x_j||^2),$$

whence the result.

4.14 <u>Lemma</u>. Suppose that $A \in HM(H)$, that A^2 exists, and that $x \in \ell_0(H)$ with $P_n x = x$. Then

$$||P_n Ax||^2 \le ||x||.|| P_n A^2 x||.$$

<u>Proof</u>. In the sufficiency part of Lemma 4.12 we proved that $||Ax||^2 = \langle x,A^*Ax\rangle$. But $||P_n Ax||^2 \le ||Ax||^2$ and

$$\langle x,A^*Ax\rangle = \langle x,A^2 x\rangle$$

$$= \langle x,P_n A^2 x\rangle$$

$$\le ||x||.||P_n A^2 x||,$$

which yields the result.

We first characterize the Hermitian matrices in (D,D):

4.15 Theorem. Let $A \in HM(H)$. Then $A \in (D,D)$ if and only if

(4.35) A^n exists for all $n \in N$,

(4.36) $K = \sup\{||(A^n)_{ii}||^{1/n} : i,n \in N\} < \infty.$

Furthermore $||A|| = K$.

Proof. Necessity. Since (D,D) is an algebra we must have (4.35). Now if $A \in (D,D)$, and $i, n \in N$, then

$$||(A^n)_{ii}|| \le ||A^n|| \le ||A||^n,$$

whence (4.36) holds and $K \le ||A||$.

Sufficiency. Let (4.35), (4.36) hold. We first prove that

(4.37) $K = \sup\{||(A^n)_{ij}||^{1/n} : i,j,n \in N\}.$

Take $i, j, n \in N$ and $x_j \in H_j$. Then, writing $B = A^n$,

$$\begin{aligned} ||B_{ij}x_j||^2 &= \langle B_{ij}x_j, B_{ij}x_j\rangle \\ &\le \sum_{k=1}^{\infty} \langle B_{kj}x_j, B_{kj}x_j\rangle \\ &= \langle \sum_{k=1}^{\infty} B^*_{kj}B_{kj}x_j, x_j\rangle \\ &= \langle (B^*B)_{jj}x_j, x_j\rangle \\ &= \langle (B^2)_{jj}x_j, x_j\rangle \end{aligned}$$

$$\leq ||(B^2)_{jj}||\ ||x_j||^2,$$

whence $||B_{ij}||^2 \leq ||(B^2)_{ij}|| \leq K^{2n}$. Thus the right side of (4.37) does not exceed K. It is trivial that K does not exceed the right side of (4.37).

Now let $x \in \ell_0(H)$, $P_n x = x$, and $||x|| \leq 1$. By Lemma 4.14,

$$||P_n Ax||^4 \leq ||x||^2 ||P_n A^2 x||^2 \leq ||P_n A^4 x||,$$

and proceeding inductively we find, writing $q = 2^r$,

$$||P_n Ax|| \leq ||P_n A^q x||^{1/q}$$

$$= ||P_n A^q P_n x||^{1/q}$$

$$\leq n^{1/q} \max\{||(A^q)_{ij}||^{1/q} : i,j \leq n\}$$

$$\leq Kn^{1/q},$$

on using Lemma 4.13 and (4.37).

Letting $r \to \infty$, we obtain $||P_n Ax|| \leq K$. But this last inequality holds for all sufficiently large n, and so $||Ax|| \leq K$. Thus A defines a bounded linear operator on $\ell_0(H)$ into D and the norm of A does not exceed K.

As in the proof of Lemma 4.12 we find that $A \in (D,D)$, and since $\ell_0(H)$ is dense in D it follows that $||A|| \leq K$.

Finally, from the necessity part of the proof we see that $||A|| = K$.

4.16 Theorem. *Let* $A \in M(H)$. *Then* $A \in (D,D)$ *if and only if*

(4.38) $\qquad (A^*A)^n$ *exists for all* $n \in N$,

(4.39) $\qquad K = \sup\{||[(A^*A)^n]_{ii}||^{1/n} : i, n \in N\} < \infty.$

<u>Furthermore</u> $||A|| = K^{1/2}$.

<u>Proof</u>. This follows from Lemma 4.12, Theorem 4.15 and the identity $||A||^2 = ||A^*A||$.

We remark that the proof of Theorem 4.15 shows that we may replace n in (4.39) by 2^k. Hence, when $A \in (D,D)$ we have

$$||A||^2 = \sup\{||[(A^*A)^{2^k}]_{ii}||^{2^{-k}} : i \geq 1,\ k \geq 0\}.$$

For some results on the compactness of matrices in (D,D) we refer the reader to Maddox and Wickstead [46]. In particular, it is shown in [46] that the (C,1) matrix does not act compactly on (ℓ_2,ℓ_2). The result that the (C,1) matrix is in (ℓ_2, ℓ_2) is essentially due to Hardy; see for example [21], p.239.

The above work on (D,D) depends heavily on the Hilbert space setting. More generally, one would like to see a solution to the following problem:

4.17 <u>Problem</u>. <u>Let</u> X, Y <u>be Banach spaces and</u> $A_{nk} \in B(X,Y)$. <u>Determine necessary and sufficient conditions for the matrix</u> (A_{nk}) <u>to be in</u> $(\ell_2(X), \ell_2(Y))$ <u>in terms of the</u> A_{nk} <u>alone</u>.

We conclude this section with an operator version of Kuttner's theorem [29]. Kuttner's result asserts that if A is a scalar Toeplitz matrix, and $0 < p < 1$, then there is a sequence summable w_p but not summable A.

In Theorms 4.18 and 4.19 we suppose that X and Y are Banach spaces and that $A_{nk} \in B(X,Y)$.

Our operator version of Kuttner's theorem depends on the characterization of the class $(w_p(X), c(Y))$ when $0 < p < 1$. We refer the reader to Maddox [41] for the proof. Also in [41] the case $1 \leq p < \infty$ is considered.

4.18 Theorem. Let $0 < p < 1$. Then $A \in (w_p(X), c(Y))$ if and only if

(4.40) there exists $\lim_n A_{nk} = A_k$ for each k,

(4.41) $$M_1 = \sup \sum_{r=0}^{\infty} 2^{r/p} \max_r ||A^*_{nk} f|| < \infty,$$

(4.42) $$M_2 = \sup \sum_{r=0}^{\infty} 2^{r/p} \max_r ||(A^*_{nk} - A^*_k) f|| < \infty,$$

where in (4.41), (4.42) the supremum is over all $n \in N$ and all $f \in S^*$. Also, $\max_r$ is over $2^r \le k < 2^{r+1}$.

4.19 Theorem. Let $0 < p < 1$ and let T be a fixed nonzero element of $B(X,Y)$. Suppose that $A \in (c(X), c(Y))$ and

(4.43) $$\lim_n \Sigma A_{nk} x_k = T(\lim x_n)$$

for all $x \in c(X)$. Then there is a sequence in $w_p(X)$ which is not summable A.

Proof. Suppose, if possible, that $A \in (w_p(X), c(Y))$. Take $z \in X$ such that $T(z) \neq \Theta$. Then (4.43) implies

(4.44) $$\lim_n \Sigma A_{nk} z = T(z),$$

and $\lim_n A_{nk} = 0$ for each k. Hence (4.41) and (4.42) reduce to the same condition.

Write $R_{nm} = (A_{nm}, A_{n,m+1}, \ldots)$. Take $n \in N$, $m \ge 2^i$ and for each $\varepsilon > 0$ choose

$$2^{i/q} < \varepsilon, \text{ where } q = p/(p-1).$$

Then, as in Proposition 3.11, $||R_{nm}|| \le \varepsilon M_1$ and so

$$\lim_n \sup_n ||R_{nm}|| = 0.$$

Thus the conditions of Theorem 4.8 hold, and so $A \in (\ell_\infty(X), c_0(Y))$. Consequently, (4.44) implies $T(z) = \theta$, which is contrary to the choice of z. This proves the theorem.

5. Tauberian theorems

In 1897, Tauber [73], with some analytically simple results on Abel summability, initiated the far-reaching subject of what is now called Tauberian theory. This theory has attracted the attention of many famous mathematicians - Hardy, Littlewood, Karamata, Ingham, Schmidt, and Wiener, to name but a few. An extensive account of the classical theory may be found in Hardy's book [19].

As a general rule, a Tauberian theorem may be described as one in which the summability to s of a series Σa_n by a summability method A, together with a restriction on (a_n) implies that Σa_n converges to s. The restriction on (a_n), or Tauberian condition, may take various forms, but is usually a condition which restricts the 'size' of a_n, such as

(5.1) $$na_n = o(1), \text{ i.e. } na_n \to o \ (n \to \infty),$$

or

(5.2) $$na_n = O(1), \text{ i.e. } \sup n|a_n| < \infty,$$

or

(5.3) $$\sum_{k=0}^{n} ka_k = o(n).$$

Below we list four of the classical Tauberian theorems. The series involved are series of complex terms.

5.1 Theorem (Tauber's first theorem). *If $\Sigma a_n = s$ (Abel) and (5.1) holds then Σa_n converges to s.*

5.2 Theorem. (Tauber's second theorem) *If $\Sigma a_n = s$ (Abel) then Σa_n converges to s if and only if (5.3) holds.*

5.3 Theorem (Hardy). *If $\Sigma a_n = s(C,\alpha)$ and (5.2) holds then Σa_n converges to s.*

5.4 Theorem (Littlewood). *If* $\Sigma a_n = s$ (Abel) *and* (5.2) *holds then* Σa_n *converges to* s.

The most significant of the above results, though not the deepest, is that of Hardy [18], in which he replaced the little o conditions of Tauber by a big O condition. In [18] Hardy raised the question as to whether (C,α) summability in Theorem 5.3 could be replaced by Abel summability. Recall that for $\alpha > -1$, if $\Sigma a_n = s$ (C,α) then $\Sigma a_n = s$ (Abel). Hardy stated that he was inclined to think that (C,α) could not be replaced by Abel summability. However, Littlewood in his famous paper [35] proved Theorem 5.4, the deepest of the above Tauberian results, which gave tremendous impetus to the whole subject of Tauberian theory, leading to the very general work of Wiener, [77], [78].

From our viewpoint the interest is in slightly more abstract Tauberian theorems which are in the setting of Banach or more general spaces.

Littlewood's Theorem 5.4 was placed in a Banach space context by Northcott [59], and we shall give an account of this generalization. First we extend Theorem 5.3, in the case $\alpha = 1$, to Banach spaces.

5.5 Theorem. *Let* Σa_n *be a series in a Banach space. If* $\Sigma a_n = s(C,1)$ *and*

$$\sup n\|a_n\| < \infty \tag{5.4}$$

then Σa_n *converges to* s.

Proof. We may take $s = \Theta$; otherwise consider Σb_n where $b_0 = a_0 - s$, $b_n = a_n$ $(n \geq 1)$. Write

$$s_n = a_0 + a_1 + \ldots + a_n,$$

$$t_n = \frac{1}{n+1} \sum_{k=0}^{n} s_k,$$

so that $\Sigma a_n = \Theta(C,1)$ is equivalent to the assertion that $||t_n|| \to 0$ $(n \to \infty)$.

Denote the supremum in (5.4) by M. Then for $q + 1 < n$,

$$(n+1)\, t_n = \sum_{0}^{q} s_k + \sum_{q+1}^{n} (s_k - s_n + s_n)$$

$$= (q+1)t_q + (n-q)s_n - \sum_{q+1}^{n} (a_{k+1} + \ldots + a_n).$$

Hence $(n-q)||s_n||$ is less than or equal to

$$(q+1)||t_q|| + (n+1)||t_n|| + M \sum_{q+1}^{n} \left(\frac{1}{k+1} + \ldots + \frac{1}{n}\right)$$

$$\le (q+1)||t_q|| + (n+1)||t_n|| + \frac{M}{q+1} \sum_{q+1}^{n} (n-k)$$

$$\le (q+1)||t_q|| + (n+1)||t_n|| + \frac{M(n-q)^2}{2(q+1)} .$$

Now choose $0 < \varepsilon < 1/2$ and $q = n-[\varepsilon n]$, where the square brackets denote the integer part. Then for all sufficiently large n,

$$||s_n|| \le \frac{4}{\varepsilon} (||t_q|| + ||t_n||) + \frac{M\varepsilon}{2(1-\varepsilon)} .$$

Since $1/2(1-\varepsilon) < 1$, there is a positive constant H such that $||s_n|| \le H\varepsilon$ for all sufficiently large n. This proves the theorem.

An interesting consequence of Theorem 5.5 for real series is that Dirichlet's theorem on the convergence of a Fourier series follows from Fejér's theorem on (C,1) summability, since $f \in BV$ implies $na_n = O(1)$ and $nb_n = O(1)$, where a_n and b_n are the Fourier coefficients of f.

5.6 Theorem. Let Σa_n be a series in a Banach space. If $\Sigma a_n = s$ (Abel) and condition (5.4) holds then Σa_n converges to s.

<u>Proof</u>. $\Sigma a_n = s$ (Abel) means that $f(x) = \sum_{k=0}^{\infty} a_k x^k$ converges in norm for each $x \in (0,1)$ and $f(x) \to s$ $(x \to 1-)$. By altering the first term of Σa_n we may without loss of generality assume that $s = \Theta$. Denote the supremum in (5.4) by M.

If $0 < x < 1$ then $||a_k x^k|| \le x^k M/k \le Mx^k$ for $k \ge 1$, whence $\Sigma a_k x^k$ is absolutely convergent. Now write

$$s_n = a_0 + a_1 + \ldots + a_n.$$

Then for $0 < x < 1$ and $n > 1$,

$$||s_n|| \le ||\sum_{k=1}^{n} (1-x^k) a_k|| + \sum_{k=n+1}^{\infty} ||a_k|| x^k + ||f(x)||$$

$$\le (1-x) \sum_{k=1}^{n} k||a_k|| + \frac{M}{n} \sum_{k=n+1}^{\infty} x^k + ||f(x)||$$

$$\le (1-x) \sum_{k=1}^{n} k||a_k|| + \frac{M}{n(1-x)} + ||f(x)||.$$

Letting $x = 1 - n^{-1}$ and using the fact that $f(1-n^{-1}) \to \Theta$ we see that

$$H = \sup||s_n|| < \infty. \tag{5.5}$$

By partial summation,

$$\sum_{k=0}^{m} a_k x^k = s_m x^m + \sum_{k=0}^{m-1} s_k (x^k - x^{k+1}),$$

so $0 < x < 1$ and (5.5) imply

$$\Sigma a_k x^k = (1-x) \Sigma s_k x^k \to \Theta \quad (x \to 1-) \tag{5.6}$$

We shall shortly show that (5.5) and (5.6) imply

$$s_n \to \Theta \ (C,1). \tag{5.7}$$

Then by Theorem 5.5, the condition (5.4), and (5.7) imply $\Sigma a_k = \Theta$, which is our result.

To prove (5.7) define a function g by

$$g(t) = 0 \text{ for } 0 \le t < e^{-1},$$

$$g(t) = t^{-1} \text{ for } e^{-1} \le t \le 1.$$

For n > 1 let $x = e^{-1/n}$, so $0 \le k \le n$ implies

$$g(x^k) = g(e^{-k/n}) = e^{k/n},$$

and $k > n$ implies $g(x^k) = 0$. Then

$$(1-x) \sum_{k=0}^{\infty} x^k g(x^k) s_k = (1-e^{-1/n}) \sum_{k=0}^{n} s_k.$$

Since $(n+1)(1-e^{-1/n}) \to 1$ $(n \to \infty)$, we see that (5.7) will follow if we prove that

$$G(x) = (1-x) \sum_{k=0}^{\infty} x^k g(x^k) s_k \to \Theta \quad (x \to 1-). \tag{5.8}$$

To do this take any polynomial p and write $h = g-p$, so that

$$G(x) = (1-x)\Sigma x^k h(x^k) s_k + (1-x)\Sigma x^k p(x^k) s_k. \tag{5.9}$$

Now for each $r \in N$, $(1-x)/(1-x^r) \to 1/r$ $(x \to 1-)$, so that (5.6) implies the second sum in (5.9) tends to Θ as $x \to 1-$.

By (5.5) the norm of the first sum in (5.9) does not exceed

$$H.(1-x)\Sigma x^k |h(x^k)|.$$

Take any $\varepsilon > 0$. By the Weierstrass approximation theorem we may choose a polynomial p such that

$$\int_0^1 |h(t)| dt < \varepsilon.$$

It is sufficient, to complete the proof, to show that

$$\limsup_{x \to 1-} (1-x)\Sigma x^k |h(x^k)| \le \int_0^1 |h(t)| dt, \tag{5.10}$$

or making a change of variable,

(5.11) $$\limsup_{y\to 0+} A \le \int_0^1 |h(t)|\,dt,$$

where

(5.12) $$A = y\Sigma e^{-ky}|h(e^{-ky})|.$$

To prove (5.11) take $0 < y < 1$, $w > 3$, and write

$$\alpha = [1/wy],\ \beta = [w/y],$$

where the square brackets denote the integer part. Now write

(5.13) $$A = A_1 + A_2 + A_3,$$

where A_1 is the sum over $0 \le k \le \alpha$, A_2 is the sum over $\alpha + 1 \le k \le \beta$, and A_3 is the sum over $k > \beta$.

By the construction of h there exists $D > 0$ such that $|h(t)| \le D$ for $0 \le t \le 1$, whence

$$A_1 \le yD(\alpha+1) \le yD + (D/w),$$

(5.14) $$\limsup_{y\to 0+} A_1 \le D/w.$$

Also,

$$A_3 \le yDe^{-w/2} \sum_{k=\beta+1}^{\infty} e^{-ky/2}$$

$$\le yDe^{-w/2}/(1-e^{-y/2}),$$

(5.15) $$\limsup_{y\to 0+} A_3 \le 2De^{-w/2}.$$

Since $\alpha y \to 1/w$ and $\beta y \to w$ as $y \to 0+$ we see that

$$(5.16) \qquad \limsup_{y\to 0+} A_2 \le \int_{1/w}^{w} e^{-t}|h(e^{-t})|dt.$$

Consequently, by (5.12) to (5.16), on letting $w \to \infty$,

$$\limsup_{y\to 0+} A \le \int_0^\infty e^{-t}|h(e^{-t})|dt = \int_0^1 |h(t)|dt,$$

whence (5.11) holds, and the proof is complete.

We next consider generalizations of Theorem 5.2 (Tauber's second theorem). These generalizations appear in Maddox [42]. There is connected work, though restricted to complex sequences, in Meyer-König and Tietz [56], [57], [58].

Let (q_n) be a sequence of complex numbers such that $Q_n = q_0 + q_1 + \ldots + q_n$ is non-zero for all $n \ge 0$. Write

$$a_{nk} = q_k/Q_n \text{ for } 0 \le k \le n,$$

$$a_{nk} = 0 \qquad \text{for } k > n.$$

Then (a_{nk}) is a Toeplitz matrix, i.e. satisfies the conditions of Theorem 2.9 if and only if

$$(5.17) \qquad |Q_n| \to \infty \text{ and } \sum_{k=0}^{n} |q_k| = O(|Q_n|).$$

In what follows we suppose that (5.17) holds. Also, we write

$$p = (p_k) = (p_1, p_2, \ldots) = (q_k/Q_{k-1}).$$

Now suppose that X is a Hausdorff topological linear space with zero θ and let $c(X)$ denote the set of all convergent sequences $x = (x_k)$ in X. For a series Σa_k of elements of X we define

$$t_n = \frac{1}{Q_n} \sum_{k=1}^{n} Q_{k-1} a_k.$$

We consider three types of Tauberian condition:

(5.18) $$a_n = (m + \varepsilon_n)p_n,$$

(5.19) $$t_n \to \Theta,$$

(5.20) $$t_n \to m.$$

In (5.18) we denote by (ε_n) a sequence in X such that $\varepsilon_n \to \Theta$ $(n \to \infty)$, and m is an element of X.

If (λ_n) is a real sequence such that $0 = \lambda_0 < \lambda_1 < \ldots < \lambda_n \to \infty$ and we put $q_k = \lambda_{k+1} - \lambda_k$ then (5.18) with $m = \Theta$, and (5.19) become the usual conditions employed in the theory of Dirichlet's series - see e.g. Hardy and Riesz [22]. In particular, if $\lambda_n = n$ then we obtain the classical Tauberian conditions (5.1) and (5.3).

For our present purpose we shall define a summability method A, not necessarily given by a matrix, to be a function

$$A : S(X) \to X$$

where S(X) is some superset of c(x). We say that A is <u>regular</u> if Ax = lim x for every x in c(X). We call A <u>additive</u> if, whenever x, y $\in$ S(X) then x + y $\in$ S(X) and A(x+y) = Ax + Ay.

A series Σa_k in X is said to be summable A to s, written $\Sigma a_k = s(A)$ if

$$(a_1 + \ldots + a_n) \in S(X) \text{ and } A((a_1 + \ldots + a_n)) = s.$$

A restriction on (a_n) is called a Tauberian condition for A if $\Sigma a_k = s(A)$, together with the restriction on (a_n) implies that Σa_k converges to s. For example, (5.3) is a Tauberian condition for the Abel method.

5.7 <u>Theorem</u>. <u>Let</u> A <u>be a regular additive summability method</u>.

(i) <u>If (5.18) is a Tauberian condition for</u> A <u>then so is</u> $(t_n) \in c(X)$.

(ii) <u>If</u> X <u>is locally convex, and</u> $(t_n) \in c(X)$ <u>is a Tauberian condition for</u> A <u>then so is (5.18)</u>.

<u>Proof</u>. (i) Suppose that (5.20) holds and that $\Sigma a_k = s(A)$. Now $a_1 = (Q_1/q_0)t_1$ and for $n \geq 2$,

$$a_n = (Q_n/Q_{n-1})t_n - t_{n-1} = (1+p_n)t_n - t_{n-1}.$$

Hence

(5.21) $$a_1 + \ldots + a_n = t_n + p_1 t_1 + \ldots + p_n t_n.$$

By hypothesis we have $A((a_1+\ldots+a_n)) = s$, and since $t_n \to m$, the regularity of A implies that $A(t) = m$.

Since $-t$ is also convergent and $A(t+(-t)) = \lim \Theta = \Theta$ it follows from (5.21) by additivity that

(5.22) $$\Sigma p_k t_k = (s-m)(A).$$

But $p_k t_k = p_k (m+\varepsilon_k)$, and since (5.18) is a Tauberian condition for A it follows from (5.22) that $\Sigma p_k t_k$ converges with sum $s-m$. By (5.21) we now see that

$$a_1 + \ldots + a_n \to m + s - m \ (n \to \infty),$$

whence Σa_k converges to s.

(ii) Let (5.18) hold and $\Sigma a_k = s(A)$. Then

$$t_n = m + \frac{1}{Q_n} \sum_{k=0}^{n} q_k \varepsilon_k,$$

if we define $\varepsilon_0 = -m$. To prove our result it is enough to show that $t_n \to m$ $(n \to \infty)$, or $\Sigma a_{nk}\varepsilon_k \to \Theta$ $(n \to \infty)$, where a_{nk} is defined just prior to (5.17). Since (5.17) is assumed to hold there is a positive constant H such that

(5.23) $$\sup_n \Sigma|a_{nk}|H^{-1} \le 1.$$

Let $N(\Theta)$ be any neighbourhood of Θ. Then $H^{-1}N(\Theta)$ is a neighbourhood of Θ, and since X is locally convex there is an absolutely convex neighbourhood U of Θ contained in $H^{-1}N(\Theta)$. Also $2^{-1}U$ is an absolutely convex neighbourhood of Θ. Since $\varepsilon_k \to \Theta$ $(k \to \infty)$ there exists $i \in N$ such that

(5.24) $$\varepsilon_k \in 2^{-1}U, \text{ for all } k > i.$$

By (5.23), (5.24),

$$H^{-1} \sum_{k>i} a_{nk}\varepsilon_k \in 2^{-1}U$$

for $n \in N$. Now $a_{nk} \to 0$ $(n \to \infty)$ for each k and so there exists $j \in N$ such that

$$H^{-1} \sum_{k\le i} a_{nk}\varepsilon_k \in 2^{-1}U$$

for all $n > j$. Hence the absolute convexity of $2^{-1}U$ implies

$$2^{-1}H^{-1}\Sigma a_{nk}\varepsilon_k \in 2^{-1}U$$

for all $n > j$, and so $\Sigma a_{nk}\varepsilon_k \in N(\Theta)$ for all $n > j$, which proves the theorem.

It was remarked by Maddox [42] that it seemed unlikely that, in general, if $a_n = \varepsilon_n p_n$ is a Tauberian condition for A then so also is (5.20). This point was settled by Kuttner [31], who applied a theorem of Darevsky [12] on methods which sum 'essentially only' a given divergent sequence. In view of the intrinsic interest of Darevsky's theorem we combine the special case of his theorem that is needed, with the argument of Kuttner:

5.8 <u>Theorem.</u> <u>There is a regular summability method given by a matrix</u> A <u>such that</u> $na_n \to 0$ <u>is a Tauberian condition for</u> A <u>but</u> $na_n \to m$ <u>is not</u>.

<u>Proof.</u> Define $b_n = 1/n$ for $n \in N$, so that

$$B_n = \sum_{k=1}^{n} b_k \to \infty \ (n \to \infty).$$

Let $n_1 = 1$ and choose $1 < n_2 < n_3 < \ldots$ such that for $k \geq 1$,

(5.25) $$kB_n \leq B_{n_{k+1}} \text{ for all } n \leq n_k.$$

Define a matrix transformation A by

(5.26) $$A_n = A_n(x) = x_n - B_n x_{n_{k+1}} / B_{n_{k+1}}$$

for $n_{k-1} < n \leq n_k$ when $k > 1$, and for $n = 1$ when $k = 1$. It follows from (5.25) that A is regular. Also $A_n(B) = 0$, so that $\Sigma b_k = 0(A)$. But $nb_n = 1$ and Σb_n diverges.

We now show that $na_n \to 0$ and $\Sigma a_n = s(A)$ imply Σa_n converges to s.

Since $\Sigma a_n = s(A)$ we have $A_n(x) \to s$ $(n \to \infty)$ with $x_n = a_1 + a_2 + \ldots a_n$.

Write $y_k = x_{n_k}$, $\alpha_k = A_{n_k}$, $\beta_k = B_{n_k}$. Then with $n = n_k$ in (5.26),

$$\alpha_k/\beta_k = y_k/\beta_k - y_{k+1}/\beta_{k+1},$$

$$y_{i+1}/\beta_{i+1} = y_1/\beta_1 - \sum_{k=1}^{i} (\alpha_k/\beta_k).$$

It follows from (5.25) that $\Sigma(1/\beta_k) < \infty$, whence

(5.27) $$y_{i+1} = \lambda\beta_{i+1} + \beta_{i+1} \sum_{k=i+1}^{\infty} (\alpha_k/\beta_k)$$

for some constant λ.

Since $\alpha_k \to s$ it is clear from (5.25) that the second term in (5.27) is a Toeplitz transformation of the sequence (α_k) and so $y_{i+1} = \lambda\beta_{i+1} + s + o(1)$. Now (5.26) implies

(5.28) $$x_n = A_n + \lambda B_n + o(1) = s + \lambda B_n + o(1),$$

and so

$$\frac{1}{B_n}\sum_{k=1}^{n} a_k = \lambda + o(1). \tag{5.29}$$

But the left hand side of (5.29) is clearly a Toeplitz transformation of the null sequence (ka_k), so (5.29) implies $\lambda = 0$. Thus (5.28) yields $x_n = s + o(1)$, which means that Σa_k converges to s, which completes the proof.

It is clear why the matrix A in Theorem 5.8 is said to sum 'essentially only' the divergent sequence B. For A sums B, and if x is summed by B to s then (5.28) shows that x must be of the form $\lambda B + t$, where t is convergent to s.

Kuttner [31] points out that by using a result of Zeller [85] one may also ensure that a normal matrix A can be used in Theorem 5.8. Recall that A is normal if $a_{nk} = 0$ for $k > n$, and $a_{nn} \neq 0$ for all n.

In the present context the following proposition, though trivial, is relevant.

5.9 Proposition. $na_n = O(1)$ *is a Tauberian condition for the* (C,1) *method but*

$$\frac{1}{n+1}\sum_{k=1}^{n} ka_k = O(1)$$

is not.

Proof. By Theorem 5.3, $na_n = O(1)$ is a Tauberian condition for (C,1).

However, if $a_k = (-1)^{k+1}$ then for all $n \in N$,

$$\left|\sum_{k=1}^{n} ka_k\right| < (n+1)$$

and $\Sigma a_k = \frac{1}{2}(C,1)$, but Σa_k diverges.

Next we show that there is quite a wide class of methods for which (5.20) is a Tauberian condition whenever $a_n = \varepsilon_n p_n$ is a Tauberian condition.

Suppose X is a Banach space and that $G_{nk} \in B(X,X)$ for $n,k \in N$. Consider the series to sequence summability method G defined by the transformation

$$G_n(a) = \Sigma G_{nk} a_k.$$

Write $\Delta G_{nk} = G_{nk} - G_{n,k+1}$. As usual, $\Sigma a_k = s(G)$ means that $G_n(a) \to s$ $(n \to \infty)$. Also, if $\sup_n \|G_n(a)\| < \infty$ then we say that Σa_k is G bounded.

We shall refer to the following conditions:

(5.30) $$\sup_n \|(\Delta G_{nk})_{k \geq 1}\| < \infty,$$

(5.31) $$G_{nk} \to I \ (n \to \infty, \text{ each } k),$$

(5.32) $$g_n \equiv \|(p_k G_{nk})_{k \geq 1}\| < \infty \ (\text{each } n),$$

(5.33) $$\lim \sup_n g_n^{-1} \|\Sigma p_k G_{nk} x\| > 0 \ (\text{each } x \neq \Theta).$$

In (5.31), I denotes the identity operator, and in (5.32) we assume that $\Sigma p_k G_{nk}$ converges for each n.

Before the main result (Theorem 5.11) we give a lemma.

5.10 _Lemma_. _Let_ (5.30) - (5.33) _hold_. _Suppose that_ Σb_k _is a series in the Banach space_ X _such that_ Σb_k _is_ G _bounded and such that_ $b_k = (m+\varepsilon_n)p_k$, _where_ $m \in X$ _and_ $\varepsilon_k \in X$, _with_ $\|\varepsilon_k\| \to 0$ $(k \to \infty)$. _Then_ $m = \Theta$.

Proof. Suppose, if possible, that $m \neq \Theta$. Now

$$\sum_{k=1}^{r} G_{nk} b_k = \sum_{k=1}^{r} p_k G_{nk} m + \sum_{k=1}^{r} p_k G_{nk} \varepsilon_k.$$

The series $\Sigma p_k G_{nk}$ converges by assumption, and it follows from (5.32), since $\|\varepsilon_k\| \to 0$ that $\Sigma p_k G_{nk} \varepsilon_k$ converges for each n. Hence

(5.34) $$\Sigma p_k G_{nk} m = \Sigma G_{nk} b_k - \Sigma p_k G_{nk} \varepsilon_k.$$

We now show that $g_n \to \infty$ $(n \to \infty)$. Let A be a positive number. Since

$1 + p_k = Q_k/Q_{k-1}$ we see that the infinite product

$$\prod_{k=1}^{\infty} (1+p_k)$$

diverges, whence $\Sigma|p_k| = \infty$, so choose r such that

$$\sum_{k=1}^{r} |p_k| > A.$$

Define $x_k = (\operatorname{sgn} p_k)m/||m||$ for $1 \le k \le r$. Then by definition of group norm

$$g_n \ge ||\sum_{k=1}^{r} |p_k|G_{nk}m||.||m||^{-1}. \tag{5.35}$$

By (5.31) and (5.35),

$$A < \sum_{k=1}^{r} |p_k| \le \limsup_n g_n$$

so $g_n \to \infty$. For sufficiently large n let $A_{nk} = g_n^{-1} p_k G_{nk}$ so that $||(A_{nk})_{k\ge1}|| = 1$. Also, (5.31) and $g_n \to \infty$ imply $\lim_n A_{nk} = 0$ for each k. Thus (A_{nk}) is an operator matrix mapping null sequences into null sequences. From (5.34) we obtain

$$g_n^{-1}||\Sigma p_k G_{nk} m|| \le g_n^{-1}||\Sigma G_{nk} b_k|| + ||\Sigma A_{nk}\varepsilon_k||.$$

Since $\sup_n||\Sigma G_{nk}b_k|| < \infty$, $g_n \to \infty$, and $\Sigma A_{nk}\varepsilon_k \to \Theta$ we see that

$$\limsup_n g_n^{-1}||\Sigma p_k G_{nk} m|| = 0,$$

which is contrary to (5.33). This proves the lemma.

5.11 Theorem. Let (5.30) - (5.33) hold. Suppose that $a_n = \varepsilon_n p_n$, with $||\varepsilon_n|| \to 0$ is a Tauberian condition for G. Then

$$t_n = \frac{1}{Q_n} \sum_{k=1}^{n} Q_{k-1} a_k \to m$$

is a Tauberian condition for G.

Proof. By partial summation,

$$\sum_{k=1}^{r} G_{nk}a_k = G_{nr}Q_rQ_{r-1}^{-1}t_r + \sum_{k=1}^{r-1} \Delta G_{nk}t_k + \sum_{k=1}^{r-1} p_kG_{nk}t_k$$

$$= G_{nr}t_r + \sum_{k=1}^{r-1} \Delta G_{nk}t_k + \sum_{k=1}^{r} p_kG_{nk}t_k.$$

From the sufficiency part of the proof of Theorem 4.5 we have

(5.36) $$H = \sup_{n,k}||G_{nk}|| < \infty.$$

Define $b_1 = t_1$, $b_k = t_k - t_{k-1}$ for $k \geq 2$. Then Σb_k converges to m, since we are assuming $t_n \to m$.

Since

(5.37) $$\sum_{k=1}^{r} G_{nk}b_k = G_{nr}t_r + \sum_{k=1}^{r-1} \Delta G_{nk}t_k$$

it follows readily that $\Sigma p_kG_{nk}t_k$ converges for each n and

(5.38) $$\Sigma G_{nk}a_k = \Sigma G_{nk}b_k + \Sigma p_kG_{nk}t_k.$$

Now by (5.37), (5.36) and (5.30), for all n,r,

$$||\sum_{k=1}^{r} G_{nk}b_k|| \leq (H + ||(\Delta G_{nk})_{k\geq 1}||)\sup||t_k||$$

and so $\sup_n||\Sigma G_{nk}b_k|| < \infty$.

Since Σa_k is G summable it now follows from (5.38) that Σp_kt_k is G bounded. But $p_kt_k = (m+\varepsilon_k)p_k$, where $||\varepsilon_k|| \to 0$, whence $m = \theta$ by Lemma 5.10. Consequently $\Sigma b_k = \theta$.

By Theorem 4.5, $\lim_n \Sigma G_{nk}b_k = \theta$, so (5.38) implies that $\lim_n \Sigma p_kG_{nk}t_k = \lim_n\Sigma G_{nk}a_k = s$, say. But $p_kt_k = p_k\varepsilon_k$ and so $\Sigma p_kt_k = s$.

Since $a_1 + \ldots + a_n = t_n + p_1t_1 + \ldots + p_nt_n$ we now see that Σa_k converges to s, which proves the theorem.

In Theorem 5.12 below we show that the hypothesis of local convexity is essential to the truth of the conclusion in Theorem 5.7(ii). To do this we work in the sequence space

$$\ell_p = \{x = (x_k) : ||x|| = \Sigma|x_k|^p < \infty\}$$

where $0 < p < 1$. The topology in ℓ_p is given by the p-norm $||.||$, which has the properties that $||\lambda x|| = |\lambda|^p||x||$ and $||x+y|| \leq ||x|| + ||y||$, for $\lambda \in C$, $x, y \in \ell_p$. Note that ℓ_p is a Hausdorff topological linear space which is not locally convex. For if $\alpha = (1-p)/p$ and

$$x^{(k)} = (0, 0, \ldots, k^{-\alpha}, 0, 0, \ldots)$$

with $k^{-\alpha}$ in the k-position, then $x^{(k)} \to \Theta$, but

$$y_n = \frac{1}{n} \sum_{k=1}^{n} x^{(k)} \not\to \Theta,$$

since $||y_n|| \to p^{-1}$.

5.12 __Theorem.__ __Let $0 < p < 1$ and consider the space ℓ_p.__ __Then__

$$t_n = \frac{1}{n+1} \sum_{k=1}^{n} ka_k \to \Theta$$

__is a Tauberian condition for the $(C,1)$ mean but $na_n \to \Theta$ is not.__

__Proof.__ That $t_n \to \Theta$ is a Tauberian condition for the $(C,1)$ mean is immediate from the relation

$$s_n = t_n + \frac{1}{n+1} \sum_{k=1}^{n} s_k.$$

Now write $a_k = (a_{ki}) = (a_{k1}, a_{k2}, \ldots)$ and let e_k be the kth. unit vector in ℓ_p. Choose natural numbers $n_1, n_2, \ldots$, as follows. Take $n_1 > 2$ and define $a_k = \Theta$ for $k < n_1$. For $k = n_1$ define $a_{ki} = 1/n_1 (n_1 < i \leq n_1 + [n_1^p])$ and $a_{ki} = 0$ otherwise. The square brackets will denote the integer part.

For $n_1 < k \leq n_1 + [n_1^p]$ define $a_k = -e_k/n_1$. Then $||s_{n_1}|| = [n_1^p]/n_1^p$,

$$s_k = \Theta \ (k = n_1 + [n_1^p]), \text{ and } ||s_k|| \le 1 \text{ for } 1 \le k \le n_1 + [n_1^p].$$

Next choose

$$(5.39) \qquad n_2 > 2^{2/p} n_1^{1/p} + 2^{1/(1-p)} 2^q$$

where $q = (1 + p^2)/p(1-p)$. Then $n_1 + [n_1^p] < n_2 - [2^p n_2^p]$. Define

$$a_k = \Theta \ (n_1 + [n_1^p] < k \le n_2 - [2^p n_2^p],$$

$$a_k = e_k/2n_2 \ (n_2 - [2^p n_2^p] < k \le n_2).$$

Then $s_k = \Theta \ (n_1 + [n_1^p] \le k \le n_2 - [2^p n_2^p])$, $||s_{n_2}|| = [2^p n_2^p]/2^p n_2^p$ and $||s_k|| \le 1$ for $1 \le k \le n_2$. Also, $n_1 + [n_1^p] < k \le n_2$ implies $||ka_k|| \le 1/2^p$.

Now define

$$a_k = -e_{k-[2^p n_2^p]}/2n_2 \quad (n_2 < k \le n_2 + [2^p n_2^p])$$

so that $s_k = \Theta \ (k = n_2 + [2^p n_2^p])$ and $||s_k|| \le 1$ for $1 \le k \le n_2 + [2^p n_2^p]$.

Also, $n_2 < k \le n_2 + [2^p n_2^p]$ implies $||ka_k|| < 1$, by (5.39).

The method of construction is now clear. We choose $n_1 < n_2 < \dots < n_i$ and then

$$(5.40) \qquad n_{i+1} > (i+1)^{2/p} n_i^{1/p} + 2^{1/(1-p)} (i+1)^q$$

with a_k defined over ranges involving terms of the form $n_i - [i^p n_i^p]$ and $n_i + [i^p n_i^p]$. We obtain

$$(5.41) \qquad ||s_k|| \le 1 \text{ for all } k,$$

$$(5.42) \qquad s_k = \Theta \ (n_i + [i^p n_i^p] \le k \le n_{i+1} - [(i+1)^p n_{i+1}^p]),$$

(5.43) $$||s_{n_i}|| = [i^p n_i^p]/i^p n_i^p \to 1 \ (i \to \infty).$$

It is clear by (5.40) that if

$$n_i + [i^p n_i^p] < k \le n_{i+1} + [(i+1)^p n_{i+1}^p]$$

then $||ka_k|| < 2^p/(1+i)^p$, whence $ka_k \to \Theta$ $(k \to \infty)$. It follows from (5.42) and (5.43) that (s_n) diverges. It remains to show that $c_n \to \Theta$, where

$$c_n = \frac{1}{n} \sum_{k=1}^{n} s_k.$$

Take $n > n_1$ so that $n_i \le n < n_{i+1}$ for some i. For simplicity we write

$$\alpha(i) = n_i - [i^p n_i^p], \text{ and } \beta(i) = n_i + [i^p n_i^p].$$

If $n_i \le n \le \beta(i)$ we consider

$$c_n = \frac{1}{n} \sum_{1}^{\beta(i-1)} s_k + \frac{1}{n} \sum_{\alpha(i)}^{n_i} s_k + \frac{1}{n} \sum_{1+n_i}^{n} s_k$$

$$\equiv A_1 + A_2 + A_3, \text{ say.}$$

Hence (5.41) implies $||A_1|| < i n_{i-1}/n^p \le i n_{i-1}/n_i^p$. It follows from (5.40) that $||A_1|| < 1/i$.

Next, by the construction of s_k we see that

$$|| \sum_{\alpha(i)}^{n_i} s_k || = i^{-p} n_i^{-p} \sum_{r=1}^{[i^p n_i^p]} r^p < (i n_i)^{p^2}$$

so that $||A_2|| < 1/i$ by (5.40). A calculation shows that $||A_3|| < 1/i$, whence $||c_n|| < 3/i$ for $n_i \le n \le \beta(i)$.

If $\beta(i) < n \le \alpha(i+1)$ we also find that $||c_n|| < 3/i$.

Finally, suppose $\alpha(i+1) < n < n_{i+1}$. It follows from (5.40) that

$2n > n_{i+1}$ whence

$$||c_n|| < \frac{3}{i} + \frac{1}{n^p}||\sum_{1+\alpha(i+1)}^{n} s_k||$$

$$\leq \frac{3}{i} + \frac{2^p\{(i+1)n_{i+1}\}^{p^2}}{n_{i+1}^p}$$

$$< \frac{3}{i} + \frac{2^p}{i+1} .$$

Combining our estimates we see that $||c_n|| \to 0$ $(n \to \infty)$, and the theorem is proved.

Theorem 5.12 was given by Maddox [44], together with other results on p-normed spaces.

6. Consistency theorems

We consider summability methods given by infinite matrices $A = (A_{nk})$ of elements $A_{nk} \in B(X,Y)$, where X, Y are Banach spaces. The A-transform of a sequence $x = (x_k) \in s(X)$ is given formally by

$$Ax = (\Sigma A_{nk} x_k)_{n \in N}$$

where the summation, as always, is over $k \in N$. We require the series to converge strongly for each n, i.e. to converge in the norm of Y.

As in Section 2 we denote the summability field of A by (A).

The basic idea of consistency of two methods A and B is that they cannot sum the same sequence to different limits. In general we have:

6.1 Definition. *Let* E *be a subset of* s(X). *Then we say* A, B *are* E-*consistent, and write* A(E)B *if and only if*

$$x \in E \cap (A) \cap (B) \text{ implies } A\text{-lim } x = B\text{-lim } x.$$

The main cases of interest are when $E = s(X)$, in which case we say merely that A and B are *consistent*, and when $E = \ell_\infty(X)$.

We now give a few simple examples involving scalar matrices.

6.2 Proposition. *Any two Cesàro means are consistent.*

Proof. If the means are (C,α) and (C,β) where $\alpha > -1$ and $\beta > -1$ then one of them must imply the other, see e.g. Hardy [19], p.100. Thus, if $x_n \to \ell (C,\alpha)$ and $x_n \to m\ (C,\beta)$, and if $\alpha \geq \beta$ then $x_n \to m\ (C,\alpha)$, whence $\ell = m$.

6.3 Proposition. *The Abel method and* (C,α) *are consistent.*

Proof. By Hardy [19], p.108, (C,α) implies Abel, for $\alpha > -1$.

6.4 Proposition. *There exist inconsistent Toeplitz matrices.*

Proof. Define A by:

$$\begin{matrix} 1 & 0 & 0 & 0 & 0 & 0 & \cdots \\ 0 & 0 & 1 & 0 & 0 & 0 & \cdots \\ 0 & 0 & 0 & 0 & 1 & 0 & \cdots \\ \cdot & \cdot & \cdot & \cdot & \cdot & \cdot & \cdots \end{matrix}$$

and B by:

$$\begin{matrix} 0 & 1 & 0 & 0 & 0 & 0 & \cdots \\ 0 & 0 & 0 & 1 & 0 & 0 & \cdots \\ 0 & 0 & 0 & 0 & 0 & 1 & \cdots \\ \cdot & \cdot & \cdot & \cdot & \cdot & \cdot & \cdots \end{matrix}$$

Then if $x = (1, 0, 1, 0, 1, 0, \ldots)$ we see that $x_n \to 1(A)$, $x_n \to 0(B)$.

6.5 <u>Corollary</u>. <u>Consistency is not an equivalence relation</u>.

<u>Proof</u>. Let I be the unit matrix. Since A, B are Toeplitz matrices we have A(s)I and I(s)B, but Proposition 6.4 shows that A(s)B is false.

6.6 <u>Proposition</u>. <u>Commuting Toeplitz matrices are ℓ_∞-consistent</u>.

<u>Proof</u>. If $x \in \ell_\infty$, then by the absolute convergence of the series involved, $(AB)x = (BA)x = A(Bx) = B(Ax)$. Thus, if $(Ax)_n \to \ell$, $(Bx)_n \to m$ then the fact that A, B are Toeplitz matrices implies $\ell = m$.

We now pose a natural question, arising out of the observation that if B is a Toeplitz matrix then it is true that $(I) \subset (B)$ and I(s)B.

Can we replace I by a general Toeplitz matrix A? The next result shows that we cannot.

6.7 <u>Proposition</u>. <u>There exist Toeplitz matrices</u> A, B <u>with</u> $(A) \subset (B)$, <u>which are inconsistent</u>.

<u>Proof</u>. Define $A_n(x) = 2x_n - x_{n+1}$ $(n \geq 2)$, $A_1(x) = -x_2$, and

$$B_n(x) = 2x_n - (1+2^{-n-1})x_{n+1} \quad (n \geq 1).$$

For $n \geq 2$ it is easy to check that

$$B_n(x) = A_n(x) + \sum_{k=1}^{n} A_k(x)2^{-k-1},$$

whence $(A) \subset (B)$. But if $x_n = 2^n$ then $x_n \to 0(A)$ and $x_n \to -1(B)$, which proves the proposition.

Let us note however in Proposition 6.7 that although A and B are inconsistent, they are ℓ_∞-consistent, since $x \in \ell_\infty$ implies $A_n(x) - B_n(x) = O(2^{-n-1}) = o(1)$. This observation is no accident, and in fact the basic general theorem on consistency, due to Mazur and Orlicz [53], asserts that for any Toeplitz matrices A, B such that every bounded sequence summed by A is also summed by B, it is true that A and B are ℓ_∞-consistent.

The first proof of this famous theorem was given by Brudno [7]. The result was merely stated by Mazur and Orlicz in [53], though a special case was given by Banach [4], p.95. The literature subsequent to Brudno [7] indicates Mazur and Orlicz had in mind a very different approach from Brudno. A functional analytic approach to the classical bounded consistency theorem may be found in Bennett and Kalton [5]. See also Ruckle [67].

Before we give an operator version we present a classical version of the bounded consistency theorem essentially due to

Petersen [61]. The proof of Petersen seems to be very condensed, and since the details are quite technical we have felt it desirable to fill out the argument where necessary. The consistency theorem is an immediate corollary of the following:

6.8 Theorem. Let A, B be scalar Toeplitz matrices and let $x \in \ell_\infty$ be summed to different values by A and B. Then there exists $y \in \ell_\infty$ which is summed by A but not by B.

Proof. Suppose α = A-lim x $\neq$ B-lim x = β, and define $s \in \ell_\infty$ by

$$s = \frac{x - \alpha e}{\beta - \alpha}\ , \text{ where } e = (1, 1, 1, \ldots).$$

Since A, B satisfy $\Sigma a_{nk} \to 1$, $\Sigma b_{nk} \to 1$ $(n \to \infty)$ we see that A-lim s = 0, B-lim s = 1.

Let us write $||s|| = \sup_k |s_k|$, and $||A|| = \sup_n \Sigma |a_{nk}|$, both $||s||$ and $||A||$ being finite. As usual, Σ without limits is over $k \in N$. Now define $F = (F_k)$ by

$$(6.1) \qquad F = (1, 0, \tfrac{1}{2}, \tfrac{2}{2}, \tfrac{1}{2}, 0, \tfrac{1}{3}, \tfrac{2}{3}, \tfrac{3}{3}, \tfrac{2}{3}, \tfrac{1}{3}, 0, \tfrac{1}{4}, \ldots).$$

Then $k = p^2$ implies $F_k = 1$, and $k = p(p+1)$ implies $F_k = 0$. Also $0 \le F_k \le 1$ for all $k \in N$ and

$$(6.2) \qquad (|F_{k+1} - F_k|) = (1, \tfrac{1}{2}, \tfrac{1}{2}, \tfrac{1}{2}, \tfrac{1}{2}, \tfrac{1}{3}, \tfrac{1}{3}, \ldots) \in c_o.$$

Now $\Sigma|a_{1k}| < \infty$ and $\Sigma|b_{1k}| < \infty$, whence there exists $r(1) \in N$ such that

$$\sum_{k=r(1)}^{\infty} (|a_{1k}| + |b_{1k}|) < 1.$$

Since $|a_{nk}| + |b_{nk}| \to 0$ $(n \to \infty$, each $k)$, there exists $m_1 \in N$ with $m_1 > 1$, such that

$$\sum_{k=1}^{r(1)} (|a_{nk}| + |b_{nk}|) < 1, \text{ for all } n \geq m_1.$$

Also, for each n with $1 < n \leq m_1$, there exists $r(n) \in N$, with r strictly increasing on $[1, m_1]$ such that

$$\sum_{k=r(n)}^{\infty} (|a_{nk}| + |b_{nk}|) < 1.$$

Moreover, there exists $m_2 > m_1$ such that

$$\sum_{k=1}^{r(m_1)} (|a_{nk}| + |b_{nk}|) < \frac{1}{2}, \text{ for all } n \geq m_2.$$

Proceeding inductively, we determine two strictly increasing sequences $(r(k))$ and (m_k) such that for $p \geq 1$,

$$\sum_{k=r(n)}^{\infty} (|a_{nk}| + |b_{nk}|) < \frac{1}{p+1}, \text{ when } m_p < n \leq m_{p+1},$$

and

$$\sum_{k=1}^{r(m_{p-1})} (|a_{nk}| + |b_{nk}|) < \frac{1}{p}, \text{ for all } n \geq m_p.$$

In the last sum, when $p = 1$, we may define $m_o = 1$.

Now define two sequences λ and ε by

$$\lambda(n) = r(m_o), \ \varepsilon(n) = 1, \text{ for } 1 \leq n < m_2,$$

$$\lambda(n) = r(m_{p-1}), \ \varepsilon(n) = p^{-1}, \text{ for } m_p \leq n < m_{p+1}, \text{ where } p \geq 2.$$

Thus $\varepsilon(n) \to 0 \ (n \to \infty)$.

Take any $n \geq m_1$ and write

(6.3) $$\alpha(n) = \sum_{k=1}^{\lambda(n)} |a_{nk}|.$$

By the above construction we see that

(6.4) $$\alpha(n) < \varepsilon(n) \to 0 \ (n \to \infty).$$

Writing

$$\delta(n) = \sum_{k=r(n)}^{\infty} |a_{nk}|$$

we see that

(6.5) $$\delta(n) < \varepsilon(n) \to 0 \ (n \to \infty).$$

Since $n \geq m_1$ there exists $p \geq 1$ with $m_p \leq n < m_{p+1}$. Let us define $y \in \ell_\infty$ by $y_k = 0$ if $k < \lambda(m_1)$ and

(6.6) $$y_k = F_p \, s_k \quad \text{if} \quad \lambda(m_p) \leq k < r(m_p), \text{ where } p \geq 1.$$

Note that F_p is defined by (6.1), and that we also have $||y|| \leq ||s||$, both norms being ℓ_∞ norms.

Now consider $\Sigma a_{nk} y_k$, which we split into three sums Σ_1, Σ_2, Σ_3, with Σ_1 over $1 \leq k < \lambda(n)$, Σ_2 over $\lambda(n) \leq k \leq r(n)$, and Σ_3 over $k > r(n)$.

By (6.3) and (6.4),

(6.7) $$|\Sigma_1| \leq ||s|| \alpha(n) \to 0 \ (n \to \infty),$$

and by (6.5),

(6.8) $$|\Sigma_3| \leq ||s|| \delta(n) \to 0 \ (n \to \infty).$$

We shall next show that $\Sigma_2 \to 0$ $(n \to \infty)$, which will prove that A-lim y = 0.

Since $m_p \le n < m_{p+1}$ we have

$$\lambda(m_p) = \lambda(n) < \lambda(m_{p+1}) = r(m_p) \le r(n) < r(m_{p+1}).$$

Hence, splitting Σ_2 into sums over $\lambda(n) \le k < r(m_p)$, and over $r(m_p) \le k \le r(n)$, and using (6.6) we have

$$(6.9) \qquad \Sigma_2 = F_p \, \Sigma_4 \, a_{nk} s_k + (F_{p+1} - F_p) \, \Sigma_5 \, a_{nk} \, s_k,$$

where Σ_4 is over $\lambda(n) \le k \le r(n)$, and Σ_5 is over $r(m_p) \le k \le r(n)$.

If we decompose $\Sigma a_{nk} s_k$ into sums over $1 \le k < \lambda(n)$, $\lambda(n) \le k \le r(n)$, and $k > r(n)$, then by (6.3)-(6.5) and the fact that A-lim s = 0, we have

$$(6.10) \qquad |\Sigma_4| \le |\Sigma a_{nk} s_k| + ||s|| \, (\alpha(n) + \delta(n)) \to 0 \ (n \to \infty).$$

It is immediate that

$$(6.11) \qquad |\Sigma_5| \le ||s||.||A||.$$

Combining (6.7)-(6.11), and using (6.2) and the fact that $|F_p| \le 1$, we see that

$$(6.12) \qquad A\text{-}\lim y = 0,$$

so that y is summed by A.

Now write

$$B_n = \Sigma b_{nk} s_k, \text{ and } C_n = \Sigma b_{nk} y_k,$$

where we know that $B_n \to 1$, since B-lim s = 1. If we decompose C_n into sums over k, exactly as we did for $\Sigma a_{nk} y_k$, and continue to denote these by $\Sigma_1, \Sigma_2, \ldots$, except that we replace a_{nk} by b_{nk}, then

$$C_n = \Sigma_1 + \Sigma_3 + F_p \left(B_n - \sum_{k<\lambda(n)} b_{nk}s_k - \sum_{k>r(n)} b_{nk}s_k\right)$$

$$+ (F_{p+1} - F_p)\ \Sigma_5.$$

Since Σ_1, $\Sigma_3 \to 0$, $\sum_{k<\lambda(n)} b_{nk}s_k$, $\sum_{k>r(n)} b_{nk}s_k \to 0$, and since F_p and Σ_5 are bounded, we have by (6.2) that

(6.13) $$C_n = F_p + o(1).$$

Let $q \in N$. First take $p = q^2$ and $n = m_p$. Then (6.13) implies, by (6.1) that $C_n = 1 + o(1)$.

Next, take $p = q(q+1)$ and $n = m_p$. Then (6.13) implies by (6.1) that $C_n = o(1)$. It follows that (C_n) is not convergent, and so y is not summed by B, which proves the theorem.

6.9 <u>Theorem (Mazur-Orlicz bounded consistency theorem)</u>. <u>Let</u> A, B <u>be scalar Toeplitz matrices such that</u>

$$\ell_\infty \cap (A) \subset (B).$$

<u>Then</u> A <u>and</u> B <u>are</u> ℓ_∞<u>-consistent.</u>

<u>Proof</u>. If A and B are not ℓ_∞-consistent, then there exists $x \in \ell_\infty \cap (A) \cap (B)$ such that x is summed to different values by A and B. By Theorem 6.8, there exists $y \in \ell_\infty \cap (A)$ such that $y \notin (B)$, contrary to hypothesis.

6.10 <u>Definition</u>. <u>We say that</u> C <u>is</u> ℓ_∞<u>-stronger than</u> A <u>if and only if</u>

$$\ell_\infty \cap (A) \subset (C).$$

6.11 Remark. With regard to the matrices A, B of Proposition 6.4, there is no Toeplitz matrix C which is ℓ_∞-stronger than both A and B. For, with x = (1, 0, 1, 0, ...), if there was such a C, then Theorem 6.9 implies 1 = C-lim x = 0.

An interesting approach to the bounded consistency theorem is that of Zeller [83], although his result is not as strong as Theorem 6.9. He shows that if A, B are scalar Toeplitz matrices such that $(A) \subset (B)$, then A and B are ℓ_∞-consistent.

Zeller proves the important result that the summability field (A), of a conservative matrix A, is an FK-space with the seminorms

$$\sup_r \left| \sum_{k=1}^{r} a_{nk}x_k \right|, \quad n = 1, 2, \ldots,$$

$$|x_n|, \quad n = 1, 2, \ldots,$$

$$\sup_n \left| \sum_{k=1}^{\infty} a_{nk}x_k \right|.$$

Also, if (A)* is the continuous dual space of (A) then Zeller shows that $f \in (A)^*$ is of the form

$$f(x) = \Sigma\alpha_k x_k + \Sigma_n \beta_n \Sigma a_{nk}x_k + \beta \lim_n \Sigma a_{nk}x_k, \tag{6.14}$$

where $x \in (A)$, $(\alpha_k) \in \ell_1$, $(\beta_n) \in \ell_1$. These results may also be found in Wilansky [79].

It is easy to show, using the Hahn-Banach theorem and (6.14), that if $y \in \ell_\infty \cap (A)$ then $y \in \bar{c}$, where the closure is taken in the topology which makes (A) an FK-space.

Thus, there exists $x^{(r)} \in c$ such that $x^{(r)} \to y$ in (A). Now it is known that, since (A), (B) are FK-spaces with $(A) \subset (B)$, we have $x^{(r)} \to y$ in (A) implies $x^{(r)} \to y$ in (B). Since A-lim $\in (A)^*$ and B-lim $\in (B)^*$ it follows that

$$\text{(6.15)} \qquad \text{A-lim}\, x^{(r)} \to \text{A-lim}\, y, \quad \text{B-lim}\, x^{(r)} \to \text{B-lim}\, y.$$

Since A,B are Toeplitz matrices, A-lim $x^{(r)}$ = B-lim $x^{(r)}$ for each $r \in N$, and so (6.15) implies A-lim y = B-lim y, which shows that A and B are ℓ_∞-consistent.

Using Theorem 6.9, Copping [10], Theorem 3, proved the following interesting result:

6.12 Theorem. If A is a conull matrix, and B is any matrix such that $\ell_\infty \cap (A) \subset (B)$, then B is conull.

Another proof of Theorem 6.12 is given by Wilansky [80].

A simple generalization of Theorem 6.9 is:

6.13 Theorem. Let A, B be coregular and c-consistent. If $\ell_\infty \cap (A) \subset (B)$, then A and B are ℓ_∞-consistent.

Proof. By c-consistency, $a_k = \lim_n a_{nk} = \lim_n b_{nk}$ for each k, and

$$\lim_n \Sigma a_{nk} = \lim_n \Sigma b_{nk}.$$

Hence A and B have the same non-zero characteristic m, say. The result follows by applying Theorem 6.9 to the Toeplitz matrices $(a_{nk}-a_k)/m$ and $(b_{nk}-a_k)/m$.

It is shown by Chang, et al [8] that Theorem 6.13 is false for conull matrices.

We now turn to an operator version of the bounded consistency theorem which was first considered by Alexiewicz and Orlicz [3], but we first need some preliminary propositions and definitions.

6.14 Proposition. The general form for each $f \in c^*(X)$ is

$$f(x) = f(y) - \Sigma f_k(\ell) + \Sigma f_k(x_k),$$

for every $x = (x_k) \in c(X)$, where $\lim_k x_k = \ell$, $y = (\ell, \ell, \ell, \ldots)$, $f_k \in X^*$ and $\Sigma ||f_k|| < \infty$.

Proof. If $f \in c^*(X)$, and $x \in c(X)$ then

$$x-y = (x_1-\ell, \Theta, \Theta, \ldots) + (\Theta, x_2-\ell, \Theta, \Theta, \ldots) + \ldots$$

and so

$$f(x) = f(y) + \Sigma f_k(x_k-\ell),$$

where we define $f_k(z) = f(\Theta, \Theta, \ldots, z, \Theta, \Theta, \ldots)$, with z in the k-position, for each $z \in X$. Hence $f_k \in X^*$ for each k.

Take any $\varepsilon > 0$ and any $n \in N$. For each $k \leq n$ there exists $z_k \in U$ such that

$$||f_k|| < |f_k(z_k)| + \varepsilon 2^{-k}.$$

If $|f_k(z_k)| > 0$ we write $f_k(z_k) = |f_k(z_k)| \exp(i\Theta_k)$ and define $x_k = \exp(-i\Theta_k)z_k$. But if $f_k(z_k) = 0$ we define $x_k = \Theta$. Also, for $k > n$ we define $x_k = \Theta$.For this x we have $\ell = \Theta$, and so $f(y) = 0$. Hence

$$\left|\sum_{k=1}^{n} f_k(x_k)\right| \leq ||f||$$

since $||x|| \leq 1$. It follows readily that $\Sigma ||f_k|| \leq ||f|| + 2\varepsilon$, and so $\Sigma ||f_k|| \leq ||f|| < \infty$. Since $|f_k(x_k)| \leq ||f_k||\,||x||$ and $|f_k(\ell)| \leq ||f_k||\,||\ell||$, both $\Sigma f_k(x_k)$ and $\Sigma f_k(\ell)$ are absolutely convergent, whence we obtain the representation of the proposition.

6.15 Corollary. The general form for each $f \in c_o^*(X)$ is

$$f(x) = \Sigma f_k(x_k),$$

for every $x \in c_0(X)$, *where* $f_k \in X^*$ *and* $\Sigma||f_k|| < \infty$.

6.16 *Proposition*. Let $x^{(n)}$, $x \in c(X)$, *and suppose*

$$\text{(i)} \quad H = \sup_n ||x^{(n)}|| < \infty,$$

$$\text{(ii)} \quad \lim_n \lim_k x_k^{(n)} = \lim_k x_k,$$

$$\text{(iii)} \quad \lim_n x_k^{(n)} = x_k \ \text{(each } k\text{)},$$

where the limits are strong, i.e. are taken with respect to the norm of X. *Then* $x^{(n)} \to x$ $(n \to \infty)$, *weakly in* $c(X)$.

Proof. Let $f \in c^*(X)$. By Proposition 6.14, for all $z \in c(X)$,

$$f(z) = f(y) - \Sigma f_k(\ell) + \Sigma f_k(z_k),$$

where $\lim_k z_k = \ell$, $f_k \in X^*$, $\Sigma||f_k|| < \infty$. Putting $z = x^{(n)} - x$, $\ell_n = \lim_k x_k^{(n)} - \lim_k x_k$, $y^{(n)} = (\ell_n, \ell_n, \ell_n, \ldots)$, we have $|f(y^{(n)})| \le ||f|| \; ||\ell_n|| \to 0$ $(n \to \infty)$, by (ii). Also,

$$|\Sigma f_k(\ell_n)| \le \Sigma||f_k|| \; ||\ell_n|| \to 0 \ (n \to \infty).$$

Finally, if $\varepsilon > 0$, choose p such that

$$\sum_{k=p+1}^{\infty} ||f_k|| < \varepsilon.$$

Then (i) implies $|\Sigma f_k(x_k)|$ is less than or equal to

$$\sum_{k=1}^{p} ||f_k|| \; ||x_k^{(n)} - x_k|| + \sum_{k=p+1}^{\infty} ||f_k|| \; (H + ||x||),$$

so (iii) implies $\limsup_n |\Sigma f_k(z_k)| \le \varepsilon\,(H + ||x||)$, and the result follows.

6.17 *Corollary*. *Let* $x^{(n)}$, $x \in c_0(X)$. *Suppose* $\sup_n ||x^{(n)}|| < \infty$, *and* $\lim_n x_k^{(n)} = x_k$ *(each* k*), weakly in* X. *Then* $x^{(n)} \to x$ $(n \to \infty)$, *weakly in* $c_0(X)$.

We remark that the conditions in Corollary 6.17 are also necessary for $x^{(n)} \to x$ $(n \to \infty)$, weakly in $c_0(X)$.

Now we establish some concepts connected with the theory of two-norm spaces, which are required for the operator version of the bounded consistency theorem.

6.18 Definition. (i) A two-norm space $(X, ||.||, ||.||^*)$ is a triple, where X is a complex linear space with two norms.

(ii) Let $x \in X$, and (x_n) be a sequence in X. We say that (x_n) is γ-convergent to x, written $x_n \to x(\gamma)$, if and only if

$$\sup_n ||x_n|| < \infty, \text{ and } ||x_n - x||^* \to 0 \ (n \to \infty).$$

(iii) We say that $(X, ||.||, ||.||^*)$ is γ-complete if and only if every (x_n) such that $\sup_n ||x_n|| < \infty$, and

$$||x_n - x_m||^* \to 0 \ (m,n \to \infty).$$

is γ-convergent to an element of X.

(iv) We say that a set $D \subset X$ is γ-dense in X if and only if, for each $x \in X$, there exists (x_n) in D such that $x_n \to x(\gamma)$.

(v) Let f be a linear functional on X. We say that f is γ-continuous if and only if, for each $x \in X$, $x_n \to x(\gamma)$ implies $f(x_n) \to f(x)$ $(n \to \infty)$.

To make further progress we introduce two properties, which will be assumed when the need arises.

6.19 Property. $||.||$ is lower semicontinuous, in a two-norm space, with respect to γ-convergence, i.e. $x_n \to x(\gamma)$ implies $||x|| \leq \liminf_n ||x_n||$.

6.20 Property [Often called the Σ_1 postulate]. Given any $\varepsilon > 0$ and $x_o \in S = \{x \in X : ||x|| \leq 1\}$ there exists $\delta > 0$ such that any $x \in S$ satisfying $||x||^* < \delta$ is of the form $x = x_1 - x_2$ where $x_1, x_2 \in S$ and

$||x_1-x_0||^* < \varepsilon$, $||x_2-x_0||^* < \varepsilon$.

The next theorem is of vital importance; it is due to Alexiewicz [2]; see also Orlicz [60].

6.21 Theorem. Let the two-norm space $(X,||.||,||.||^*)$ be γ-complete (see Definition 6.18(iii)) and suppose Properties 6.19 and 6.20 hold. Then the pointwise limit of a convergent sequence of γ-continuous linear functionals is itself γ-continuous.

Proof. Let $S = \{x \in X : ||x|| \le 1\}$ and define $d(x_1,x_2) = ||x_1-x_2||^*$ for $x_1, x_2 \in S$. Suppose (x_n) is a Cauchy sequence in (S,d), so that $||x_n|| \le 1$, and $||x_n-x_m||^* \to 0$ $(m,n \to \infty)$. By γ-completeness there exists $x \in X$ such that $x_n \to x(\gamma)$, so by Property 6.19 we have $||x|| \le \liminf_n ||x_n|| \le 1$. Hence (S,d) is a complete metric space.

Suppose $f_n(x) \to f(x)$ $(n \to \infty)$, on X, where f_n is γ-continuous. If $x_m \to x$ $(m \to \infty)$ on (S,d) then $x_m \to x(\gamma)$ and so $f_n(x_m) \to f_n(x)$ $(m \to \infty)$, for each n. Thus, each f_n is continuous on (S,d), and $f_n(x) \to f(x)$ $(n \to \infty)$, on S.

By a result of Alexiewicz [1] it follows that the sequence (f_n) is equicontinuous at some $x_0 \in S$, so if $\alpha > 0$ is given, there exists $\varepsilon > 0$ such that $||x-x_0||^* < \varepsilon$, $x \in S$ imply $|f_n(x) - f_n(x_0)| < \alpha$, for all $n \in N$. By Property 6.20, choose $\delta > 0$, so that if $||x||^* < \delta$, $x \in S$, then $x = x_1-x_2$, $||x_i-x_0||^* < \varepsilon$, $i = 1,2$. Hence, for all $n \in N$,

$$|f_n(x)| = |f_n(x_1-x_0+x_0-x_2)| < 2\alpha,$$

and so $|f(x)| \le 2\alpha$, if $||x||^* < \delta$, $x \in S$.

Now let $x_n \to x$ (γ), with x_n, $x \in X$. Then there exists $M > 0$ such that $||x_n|| < M$ for all $n \in N$ and $||x_n-x||^* \to 0$ as $n \to \infty$. Hence $x_n/M \in S$, and $x/M \in S$. Thus, ultimately, $|f(x_n)-f(x)| \le 4M\alpha$, and so f is γ-continuous, which completes the proof.

6.22 Proposition. Suppose $A = (A_{nk}) \in (c_0(X), c_0(Y))$, i.e. A maps null sequences into null sequences. Write

$$(A)_0 = \{x \in s(X) : Ax \in c_0(Y)\},$$

fix an element $x \in \ell_\infty(X) \cap (A)_0$ and define $x^{(n)} = (x_1, x_2, \ldots, x_n, \Theta, \Theta, \Theta, \ldots)$. If $y = Ax$,

$$y^{(n)} = (\sum_{k=1}^{n} A_{jk}x_k)_{j\in N},$$

$$v_i^{(n)} = (\sum_{r=1}^{k} A_{ir}x_r^{(n)})_{k\in N}, \quad v_i = (\sum_{r=1}^{k} A_{ir}x_r)_{k\in N},$$

then

$$(y^{(n)}, v_1^{(n)}, \ldots, v_s^{(n)}) \to (y, v_1, \ldots, v_s) \ (n \to \infty),$$ weakly in the Cartesian product space $c_0(Y) \times c(Y)^{s+1}$, where $s \in N$ is fixed.

Proof. Since $M = \sup_j ||(A_{jk})_{k\in N}|| < \infty$, we have $||y^{(n)}|| \le M||x||$ for all $n \in N$. Also, $\lim_n y_j^{(n)} = y_j$ (each j), strongly in Y, whence Corollary 6.17 implies $y^{(n)} \to y$ $(n \to \infty)$, weakly in $c_0(Y)$.

Similarly, by Proposition 6.16, with Y in place of X, etc., we see that $v_i^{(n)} \to v_i$ $(n \to \infty)$, weakly in $c(Y)$, for each i, whence the result.

6.23 Theorem. Let $A \in (c_0(X), c_0(Y))$. Then $\ell_\infty(X) \cap (A)_0$ is a γ-complete two-norm space, with

$$||x|| = \sup_k ||x_k||,$$

$$||x||^* = \sup_q ||(Ax)_q|| + \Sigma 2^{-k}(||x_k|| + \sup_q ||\sum_{r=1}^{q} A_{kr}x_r||),$$

for each $x \in \ell_\infty(X) \cap (A)_0$, and Property 6.19 holds in $\ell_\infty(X) \cap (A)_0$.

Proof. It is obvious that $||x||$ and $||x||^*$ define norms for each $x \in \ell_\infty(X) \cap (A)_0$. We note that, if $M = \sup_k ||(A_{kr})_{r\in N}||$, which is finite,

then

$$\sup_q \left\| \sum_{r=1}^{q} A_{kr} x_r \right\| \le M \|x\|,$$

for each $x \in \ell_\infty(X)$ and every $k \in N$.

Now let $x^{(n)} \to x(\gamma)$. Then for each k,

$$\|x_k^{(n)} - x_k\| \le 2^k \|x^{(n)} - x\|^* \to 0 \ (n \to \infty),$$

and so $\lim_n \|x_k^{(n)}\| = \|x_k\|$. But $\|x_k^{(n)}\| \le \|x^{(n)}\|$, which implies $\|x_k\| \le \lim\inf_n \|x^{(n)}\|$, for each k, and so Property 6.19 holds.

Suppose $H = \sup_n \|x^{(n)}\| < \infty$, $\|x^{(n)} - x^{(m)}\|^* \to 0 \ (m,n \to \infty)$. Then, for each k, there exists $\lim_n x_k^{(n)} = x_k$, so that $\|x_k^{(n)}\| \le H$ implies $\|x\| \le H$, whence $x \in \ell_\infty(X)$.

For each $p > 1$,

$$\Sigma\, 2^{-k} \|x_k^{(n)} - x_k\| \le \sum^{p-1} \|x_k^{(n)} - x_k\| + 2H \sum_p^\infty 2^{-k}.$$

Applying the operator $\lim_p \lim\sup_n$ to both sides of the inequality we obtain

$$\Sigma\, 2^{-k} \|x_k^{(n)} - x_k\| \to 0 \ (n \to \infty).$$

It is striaghtforward to show that

$$\Sigma\, 2^{-k} \sup_q \left\| \sum_{r=1}^{q} A_{kr} (x_r^{(n)} - x_r) \right\| \to 0 \ (n \to \infty)$$

and

$$\sup_q \|(A(x^{(n)} - x))_q\| \to 0 \ (n \to \infty),$$

whence $\|x^{(n)} - x\|^* \to 0 \ (n \to \infty)$. Also, we see that $x \in (A)_o$ and the proof is complete.

6.24 Theorem. Let the hypotheses of Theorem 6.23 hold. Then the space $\ell_o(X)$ of finite sequences is γ-dense in the two-norm space $\ell_\infty(X) \cap (A)_o$.

Proof. Take $x \in \ell_\infty(X) \cap (A)_o$, $\varepsilon > 0$, $s > 1$, and use the notation of Proposition 6.22. Write

$$t_n = (y^{(n)}, v_1^{(n)}, \ldots, v_s^{(n)}), \text{ and } t = (y, v_1, \ldots v_s),$$

so that $t_n \to t$ $(n \to \infty)$, weakly in $c_o(Y) \times c(Y)^{s+1}$. Hence t is in the weak closure of $\{t_k : k > s\}$. By a theorem of Mazur [52] for normed spaces (see also Robertson and Robertson [65], p.34, for the extension to general locally convex spaces), it follows that t is in the norm closure of the convex hull of $\{t_k : k > s\}$. Hence there exist non-negative $\lambda_1, \ldots, \lambda_p$ with

$$\lambda_1 + \lambda_2 + \ldots + \lambda_p = 1,$$

such that

(6.16) $$||t - (\lambda_1 t_{s+1} + \ldots + \lambda_p t_{s+p})|| < \varepsilon.$$

Now write $\alpha_{s+k} = \lambda_k + \lambda_{k+1} + \ldots + \lambda_p$ for $k = 1, \ldots, p$, so that $1 = \alpha_{s+1} \geq \alpha_{s+2} \geq \ldots \geq \alpha_{s+p} \geq 0$. Define $z \in \ell_o(X)$ by

$$z = \lambda_1 x^{(s+1)} + \ldots + \lambda_p x^{(s+p)}$$

$$= (x_1, \ldots, x_s, \alpha_{s+1} x_{s+1}, \ldots, \alpha_{s+p} x_{s+p}, \Theta, \Theta, \ldots).$$

Note that $||z|| \leq ||x||$. Now choose s so large that

(6.17) $$||x|| \sum_{s+1}^{\infty} 2^{-k} < \varepsilon,$$

whence

(6.18) $$\Sigma 2^{-k} ||x_k - z_k|| < \varepsilon.$$

But

$$\lambda_1 y^{(s+1)} + \ldots + \lambda_p y^{(s+p)} = Az,$$

so (6.16) implies

(6.19) $$||A(x-z)|| < \varepsilon,$$

and for $k \le s$ we also have

$$\text{(6.20)} \qquad \sup_q \left|\left| \sum_{r=1}^{q} A_{kr}(x_r - z_r) \right|\right| < \varepsilon.$$

By (6.17) - (6.20) it follows that $||x-z||^* < (3+M)\varepsilon$, and since $||z|| \le ||x||$ our result is proved.

6.25 Theorem. Let the hypotheses of Theorem 6.23 hold. Then the space $\ell_\infty(X) \cap (A)_o$ has Property 6.20.

Proof. Let M be as in Theorem 6.23, let $\varepsilon > 0$ and take $||x^{(0)}|| \le 1$. By Theorem 6.24, there exists $z \in \ell_0(X)$ such that $||x^{(0)}-z||^* < \varepsilon/2$ and $||z|| \le 1$, with $z_k = \Theta$ for $k > p$, where we may choose p such that $2^{-p} < \varepsilon/8(M+1)$.

Now let $0 < \alpha < 1$ and $\delta = \alpha 2^{-p-1}$, where we shall later choose α small enough for our purpose.

Take $x \in S$, $||x||^* < \delta$. Hence, by definition of $||x||^*$ we see that

$$\text{(6.21)} \qquad ||x_k|| < \alpha, \text{ for } 1 \le k \le p,$$

$$\text{(6.22)} \qquad \sup_q \left|\left| \sum_{r=1}^{q} A_{kr} x_r \right|\right| < \alpha, \text{ for } 1 \le k \le p,$$

$$\text{(6.23)} \qquad \sup_q ||(Ax)_q|| < \delta.$$

Define

$$E = \{r : r \le p, ||z_r + x_r|| \le 1\},$$

$$F = \{r : r \le p, ||z_r + x_r|| > 1\},$$

$$\lambda_r = \frac{||z_r + x_r|| + \alpha - 1}{||z_r + x_r||}, \text{ for } r \in F.$$

If $r \in F$, $||z_r + x_r|| \le ||z_r|| + ||x_r|| < 1 + \alpha$, by (6.21), and so $0 < \lambda_r < 2\alpha$.

Now define

$$x_r^{(1)} = z_r + x_r; \quad x_r^{(2)} = z_r \quad (r \in E),$$

$$x_r^{(1)} = (1-\lambda_r)(z_r+x_r); \; x_r^{(2)} = z_r-\lambda_r(z_r+x_r) \quad (r \in F),$$

$$x_r^{(1)} = x_r \; ; \; x_r^{(2)} = \Theta \quad \text{(otherwise)}.$$

Then $x = x^{(1)} - x^{(2)}$, and it is easy to check that $x^{(1)}, x^{(2)} \in S$. It remains to show that

(6.24) $$||x^{(i)} - x^{(0)}||^* < \varepsilon, \text{ for } i = 1,2.$$

Now split $\Sigma_r A_{qr}(z_r-x_r^{(1)})$ into sums Σ_1 over E, Σ_2 over F, and Σ_3 over $r > p$. Then, for all $q \in N$,

$$||\Sigma_1|| \le M\alpha, \text{ by (6.21)},$$

$$||\Sigma_2|| \le M \max_F ||-x_r+\lambda_r (x_r+z_r)||$$

$$\le M(\alpha+\lambda_r\{\alpha+1\})$$

$$\le M(\alpha+2\alpha\{2\}) = 5M\alpha,$$

$$||\Sigma_3|| = ||\Sigma_{r>p} A_{qr}x_r||$$

$$= ||(Ax)_q - \sum_{r=1}^{p} A_{qr}x_r||$$

$$< \delta + M\alpha, \text{ by (6.21), (6.23)}.$$

Hence

(6.25) $$\sup_q ||(A(z-x^{(1)}))_q|| < \delta + 7M\alpha.$$

Also,

$$\Sigma_r 2^{-r}||z_r-x_r^{(1)}||$$

$$\le \sum_E 2^{-r}||x_r|| + \sum_F 2^{-r}(\alpha + \lambda_r\{\alpha+1\}) + \sum_{r>p} 2^{-r}||x_r||$$

$$\le \Sigma_r 2^{-r}||x_r|| + 5\alpha \le ||x||^* + 5\alpha$$

$$< \delta + 5\alpha.$$

In a similar way, we find the third term of the sum making up $||z-x^{(1)}||^*$

is less than $(6M + 2)\alpha + M2^{-p+1}$. Combining the results from (6.25) we obtain

$$(6.26) \qquad ||z-x^{(1)}||^* < 2\delta + 13M\alpha + 7\alpha + \frac{\varepsilon}{4} < \frac{\varepsilon}{2},$$

on choosing $\alpha = \varepsilon/(32+52M)$. Similarly, we have $||z-x^{(2)}||^* < \varepsilon/2$, so by (6.26) and the fact that $||x^{(0)} - z||^* < \varepsilon/2$ we see that (6.24) holds, and the theorem is proved.

We shall now let $wc(Y)$ denote the set of all sequences in Y which converge weakly, and $wc_0(Y)$ denote the set of all sequences in Y which converge weakly to Θ. Also,

$$B \in (c(X), wc(Y))$$

means that, for each $x \in c(X)$, and each $n \in N$,

$$(6.27) \qquad B_n(x) = \Sigma B_{nk}x_k$$

converges weakly in Y, and also $(B_n(x))_{n\in N}$ is weakly convergent in Y. By

$$B \in (c(X), wc(Y); P)$$

we mean, in addition that

$$B_n(x) \to \lim_k x_k \quad (n \to \infty), \text{ weakly in } Y,$$

for each $x \in c(X)$.

The meaning of $B \in (c_0(X), wc_0(Y))$ is now obvious. We define

$$(B)^w = \{x \in s(X) : Bx \in wc(Y)\},$$

$$(B)^w_0 = \{x \in s(X) : Bx \in wc_0(Y)\}.$$

The fundamental operator bounded consistency theorem is:

6.26 <u>Theorem</u>. <u>Let</u> $A \in (c_0(X), c_0(Y))$ <u>and</u> $B \in (c_0(X), wc_0(Y))$. <u>If</u>

$$\ell_\infty(X) \cap (A)_0 \subset (B)^w$$

then

$$\ell_\infty(X) \cap (A)_0 \subset (B)_0^w.$$

Proof. Define $B_n(x)$ by (6.27). Take any $f \in Y^*$ and define on the two-norm space $\ell_\infty(X) \cap (A)_0$,

$$f_{pn}(x) = f\left(\sum_{k=1}^{p} B_{nk}x_k\right).$$

For each p and n, f_{pn} is clearly linear.

Now let $x^{(m)} \to \Theta(\gamma)$. Then $x_k^{(m)} \to \Theta$ $(m \to \infty$, each k), by definition of $||x||^*$ in Theorem 6.23. Hence, since B_{nk} is continuous,

$$\lim_m f_{pn}(x^{(m)}) = 0,$$

so that for each p and n, f_{pn} is γ-continuous on $\ell_\infty(X) \cap (A)_0$.

For each $x \in \ell_\infty(X) \cap (A)_0$ and each n,

$$\sum_{k=1}^{p} B_{nk}x_k \to B_n(x) \ (p \to \infty), \text{ weakly in } Y,$$

so by Theorems 6.21, 6.23, 6.25 the map defined by $f(B_n(x))$ is γ-continuous on $\ell_\infty(X) \cap (A)_0$.

If $x \in \ell_\infty(X) \cap (A)_0$ then the hypothesis of the theorem gives $x \in (B)^w$, so

$$B_n(x) \to b(x), \text{ say, as } n \to \infty, \text{ weakly in } Y,$$

whence $f(b(x))$ is γ-continuous.

By Theorem 6.24, there exists $z^{(m)} \to x(\gamma)$, with $z^{(m)} \in \ell_0(X)$. Hence $B_n(z^{(m)}) \to \Theta$ $(n \to \infty)$, weakly in Y, for each m, and so $b(z^{(m)}) = \Theta$, for each m, whence $f(b(z^{(m)}) = 0$. Since $f(b(x))$ is γ-continuous it follows that $f(b(x)) = 0$. But f was arbitrary, so by a well-known result, see for example Maddox [40], p.123, Corollary 3, we must have $b(x) = \Theta$, which proves the theorem.

Our main result follows readily from the above:

6.27 Theorem (The operator bounded consistency theorem). Let A be a Toeplitz operator matrix, i.e. $A \in (c(X), c(Y); P)$ and let $B \in (c(X), wc(Y); P)$, where P denotes that limits are preserved. Suppose

$$\ell_\infty(X) \cap (A) \subset (B)^w.$$

Then A and B are $\ell_\infty(X)$-weakly consistent, in the sense that if $x \in \ell_\infty(X) \cap (A) \cap (B)^w$ then

$$B_n(x) \to A\text{-lim } x \ (n \to \infty), \text{ weakly in } Y.$$

Proof. Let $x \in \ell_\infty(X)$, $A_n(x) = \Sigma A_{nk}x_k \to \ell \ (n \to \infty)$, strongly in Y and $B_n(x) = \Sigma B_{nk}x_k \to m \ (n \to \infty)$, weakly in Y. Write $z = (x_k - \ell)$. Since $\Sigma A_{nk}\ell \to \ell \ (n \to \infty)$, strongly in Y and $\Sigma B_{nk}\ell \to \ell \ (n \to \infty)$, weakly in Y, we have $A_n(z) \to \Theta \ (n \to \infty)$, strongly in Y and

$$B_n(z) \to m - \ell \ (n \to \infty), \text{ weakly in } Y. \tag{6.28}$$

By Theorem 6.26, $\ell_\infty(X) \cap (A)_0 \subset (B)_0^w$, whence

$$B_n(z) \to \Theta \ (n \to \infty), \text{ weakly in } Y. \tag{6.29}$$

From (6.28) and (6.29) we have $\ell = m$, and the proof is complete.

7. Operator Nörlund means

We define a generalization of the usual complex Nörlund mean (see Section 2), which involves sequences of operators on a Banach space. We prove theorems concerning the equality of summability fields of generalized Nörlund means, and also prove a consistency theorem for a certain class of these generalized means. Most of the material of this section is to appear in Maddox [50].

Suppose X is a Banach space. If $(B_k) = (B_0, B_1, \ldots)$ is a sequence in $B(X,X)$ then $||(B_k)||$ denotes the group norm, as in Definition 2.1.

If $A \in B(X,X)$, then a statement such as $B_n \to A$ $(n \to \infty)$ refers to the strong operator topology, i.e. $B_n x \to Ax$ $(n \to \infty)$ for each $x \in X$.

We define a generalized (N,q,X) mean as follows:

Let $(Q_n) = (Q_0, Q_1, \ldots)$ be a sequence of invertible elements of $B(X,X)$. Define $q_0 = Q_0$ and $q_n = Q_n - Q_{n-1}$ for $n \geq 1$, so that

$$Q_n = q_0 + q_1 + \ldots + q_n.$$

If $x = (x_k)$ is a sequence in X we write

$$N_n^q(x) = \sum_{r=0}^{n} Q_n^{-1} q_{n-r} x_r. \tag{7.1}$$

We say that x is (N,q,X) summable to m, written $x_n \to m$ (N,q,X) if and only if there exists $m \in X$ such that $N_n^q(x) \to m$ $(n \to \infty)$ in the norm of X. Also, we use (N,q,X) to denote the summability field of the mean.

A mean (N,q,X) is called regular if and only if $x_n \to m$ implies $x_n \to m$ (N,q,X).

Let (N,p,X) and (N,q,X) be any means. Then the invertibility of P_0 implies the existence of a unique sequence $k = (k_n)$ of elements of $B(X,X)$ satisfying the convolution equation $k * p = q$, i.e.

$$\sum_{r=0}^{n} k_{n-r} p_r = q_n, \text{ for } n = 0, 1, \ldots .$$

It follows that $k * P = Q$. Likewise there is a unique ℓ such that $\ell * q = p$, whence $\ell * Q = P$.

In the case when $X = C$, the Banach space of complex numbers with the usual modulus norm, and Q_n may be identified with a non-zero complex number, then (N,q,C) reduces to the usual (N,q) mean.

In the general case, straightforward calculation shows that

$$(7.2) \qquad N_n^q(x) = \sum_{r=0}^{n} Q_n^{-1} k_{n-r} P_r N_r^p(x),$$

$$(7.3) \qquad N_n^p(x) = \sum_{r=0}^{n} P_n^{-1} \ell_{n-r} Q_r N_r^q(x).$$

We now prove

7.1 <u>Theorem</u>. <u>Let</u> $(P_o, Q_o, P_1, Q_1, \ldots)$ <u>be a commuting sequence of invertible operators of</u> $B(X,X)$. <u>Suppose that</u> (N,p,X) <u>is regular, and that</u> $(N,p,X) = (N,q,X)$, <u>i.e. the means have the same summability field</u>. <u>Then the sequence</u> $k = (k_n)$ <u>has finite group norm</u>.

<u>Proof</u>. Since $(N,p,X) \subseteq (N,q,X)$, it follows from Theorem 4.2 that there is an absolute positive constant H_1 such that

$$(7.4) \qquad \left|\left| \sum_{r=0}^{n} Q_n^{-1} k_{n-r} P_r y_r \right|\right| \le H_1$$

for all n and all $y_r \in S$. Likewise, there is an $H_2 > 0$ such that

$$(7.5) \qquad \left|\left| \sum_{r=0}^{n} P_n^{-1} \ell_{n-r} Q_r y_r \right|\right| \le H_2$$

for all n and all $y_r \in S$. Putting $y_r = \Theta$ for $0 \le r < n$, it follows from (7.5) that

(7.6) $$||P_n^{-1}\, \ell_o\, Q_n|| \le H_2$$

for all n. In (7.6) the norm is that of B(X,X).

By (7.4), the commutative property, and (7.6) we have

$$||\sum_{r=o}^{n} P_n^{-1} k_{n-r} P_r y_r|| \le H_1 H_2 ||Q_o P_o^{-1}||,$$

whence for any fixed r,

(7.7) $$||\sum_{i=0}^{r} k_i\, P_n^{-1}\, P_{n-i}\, y_i|| \le H, \text{ say,}$$

for all n and all $y_i \in S$.

Since (N,p,X) is regular, it follows from Robinson's theorem that $P_n^{-1}\, p_n \to 0$ $(n \to \infty)$, whence $P_n^{-1}\, P_{n-1} \to I$ $(n \to \infty)$, where I denotes the identity of B(X,X). By the Banach-Steinhaus theorem, for each fixed i,

(7.8) $$P_n^{-1} P_{n-i} \to I \ (n \to \infty).$$

Since each $k_i \in B(X,X)$ we see from (7.8) on letting $n \to \infty$ in (7.7) that

$$||\sum_{i=0}^{r} k_i y_i|| \le H,$$

which means that k has finite group norm. This proves the theorem.

We remark that the regularity of (N,p,X) implies that existence of a constant D such that

$$||\sum_{r=0}^{n} P_n^{-1}\, p_{n-r} y_r|| \le D, \text{ for all n and all } y_r \in S,$$

whence $M = \sup_n ||P_n^{-1}|| \le D||P_0^{-1}|| < \infty$. Consequently, if (7.4) holds, we have $||Q_n^{-1}\, k_O P_n|| \le H_1$ and so

(7.9) $$||Q_n^{-1} k_{n-r}|| \le MH_1 ||k_o^{-1}||\ ||k_{n-r}||,$$

for each fixed $r < n$.

7.2 Theorem. *Let* (N,p) *and* (N,q) *be complex Nörlund means with the same summability field, and with* (N,p) *regular. Then* (N,p) *and* (N,q) *are consistent.*

Proof. Taking $X = C$ in Theorem 7.1 and identifying the P_n and Q_n with non-zero complex numbers, the (finite) group norm of k is equal to

$$\sum_{i=0}^{\infty} |k_i|,$$

whence $k_{n-r} \to 0$ $(n \to \infty$, each $r)$. By (7.9),

$$Q_n^{-1} k_{n-r} \to 0 \quad (n \to \infty, \text{ each } r),$$

so by (7.4) the transformation given by (7.2) is regular. Consequently (N,p) and (N,q) are consistent.

We note that Theorem 7.2 generalizes a result of Maddox [47], COROLLARY 1, which asserts that if $c = (N,q)$ then the unit infinite matrix and (N,q) are consistent.

By examining the proof of COROLLARY 1 in [47] we see that $c = (N,q)$ implies

(7.10) $$(q_n) \in \ell_1 \text{ and } \sup_n |Q_n^{-1}| < \infty,$$

and that (7.10) implies

(7.11) (N,q) is regular.

The analogue of the assertion that (7.10) implies (7.11) for generalized Nörlund means is that (7.12) implies (7.13), where

(7.12) $$||(q_n)|| < \infty \text{ and } \sup_n ||Q_n^{-1}|| < \infty,$$

(7.13) (N,q,X) is regular.

Now if we take $p_0 = I$ and $p_n = 0$ for $n > 0$ in Theorem 7.1 so that $k_n = q_n$, then we have that $(N,p,X) = (N,q,X)$ implies (7.12). However, we now show that, in general, (7.12) does not imply (7.13). To do this we take $X = \ell_\infty$ the space of bounded sequences with the usual supremum norm. Define, for each $x = (x_k) \in \ell_\infty$,

$$Q_0 x = x,$$

$$Q_1 x = (x_0, 2x_1, x_2, x_3, \ldots),$$

$$Q_2 x = (x_0, 2x_1, 2x_2, x_3, \ldots),$$

$$Q_3 x = (x_0, 2x_1, 2x_2, 2x_3, x_4, \ldots), \ \ldots .$$

Then for each n, Q_n is in $B(\ell_\infty, \ell_\infty)$ and Q_n is invertible, with

$$Q_n^{-1} x = \left(x_0, \frac{x_1}{2}, \frac{x_2}{2}, \ldots, \frac{x_n}{2}, x_{n+1}, \ldots\right),$$

for $n \geq 1$. Hence $||Q_n^{-1}|| \leq 1$ for all n, and it is easy to check that the group norm $||(q_n)|| \leq 2$. Thus (7.12) holds, but if $x_n = 1$ for all $n \geq 0$, then $n \geq 1$ implies

$$||Q_n^{-1} q_n x|| = ||(0, 0, \ldots, \frac{1}{2}, 0, 0, \ldots|| = \frac{1}{2} .$$

Hence $Q_n^{-1} q_n \not\to 0$ $(n \to \infty)$, and so (7.13) fails.

The following theorem was proved by Maddox [47].

7.3 Theorem. Let (N,q) be a complex Nörlund mean. Then $c = (N,q)$ if and only if

$$(q_n) \in \ell_1 \text{ and } \sum_{n=0}^{\infty} q_n x^n \neq 0 \text{ on } |z| \leq 1.$$

In the next theorem we present necessary and sufficient conditions for a generalized mean (N,q,X) to have the space $c(X)$ of convergent sequences in X as its summability field. These conditions are not so elegant as those for the special case $X = C$ which are given in Theorem 7.3 above, but this would seem to be unavoidable.

7.4 Theorem. *Let* (Q_n) *be a sequence of invertible elements of* $B(X,X)$. Then $c(X) = (N,q,X)$ if and only if

(7.14) $$q \text{ and } \ell \text{ have finite group norms,}$$

(7.15) $$H = \sup||Q_n^{-1}|| < \infty,$$

(7.16) $$\text{there exists } \lim Q_n^{-1}q_n = A,$$

(7.17) $$\text{there exists } \lim \ell_n = B.$$

In (7.14) *we define* ℓ *by* $\ell * q = p$, *where* $p_o = I$ *and* $p_n = 0$ *for* $n > 0$.

Proof. Taking $p_o = I$, $p_n = 0$ for $n > 0$ we have $c(X) = (N,p,X)$. Suppose that $c(X) = (N,q,X)$. Then by (7.2) it follows that

$$\left|\left| \sum_{r=o}^{n} Q_n^{-1} q_{n-r} y_r \right|\right| \le H_1,$$

for all n and for all $y_r \in S$, and also

$$Q_n^{-1} q_{n-r} \to A_r \quad (n \to \infty, \text{ each } r).$$

Hence (7.16) holds. Also, by (7.3),

$$\left|\left| \sum_{r=o}^{n} \ell_{n-r} Q_r y_r \right|\right| \le H_2,$$

for all n and for all $y_r \in S$, and

$$\ell_{n-r} Q_r \to B_r \quad (n \to \infty, \text{ each } r).$$

Consequently $\ell_n Q_o \to B_o$, which implies (7.17). Now for all n and all $y_r \in S$,

$$||\sum_{r=o}^{n} q_{n-r}y_r|| \le ||Q_n||H_1 \le ||Q_o||H_2\, H_1$$

which implies the group norm $||(q_n)|| \le ||Q_o||H_2H_1$, so the first part of (7.14) holds.

Now choose $H_3 > H_1||Q_o^{-1}||$. Then $||Q_n^{-1}|| < H_3$ which gives (7.15) and so

$$Q_r^{-1}H_3^{-1}y_r \in S$$

whenever $y_r \in S$. Hence, for all n and all $y_r \in S$,

$$||\sum_{r=o}^{n} \ell_{n-r}y_r|| \le H_3H_2,$$

which implies the second part of (7.14). This proves the necessity.

Conversely, let (7.14) - (7.17) hold. It follows from (7.16) that $Q_n^{-1}\, q_{n-1} = I-Q_n^{-1}q_n \to I - A$, and so

(7.18) $$Q_n^{-1}q_{n-r} \to (I-A)^rA \quad (n \to \infty,\ \text{each } r),$$

and from (7.17) that

(7.19) $$\ell_{n-r}Q_r \to BQ_r \quad (n \to \infty,\ \text{each } r).$$

By (7.15) and the first part of (7.14), for all n and all $y_r \in S$,

(7.20) $$||\sum_{r=o}^{n} Q_n^{-1}\, q_{n-r}\, y_r|| \le \sup||Q_n^{-1}||.||(q_n)||.$$

Now choose $M > ||(q_n)||$. From the definition of group norm it follows that $||Q_r|| < M$ for all r. Hence, for all n and all $y_r \in S$, it follows from the second part of (7.14) that

(7.21) $$||\sum_{r=o}^{n} \ell_{n-r}Q_ry_r|| \le M||(\ell_n)||.$$

But the conditions (7.18) - (7.21) are sufficient for $c(X) = (N,q,X)$.

Next we prove:

7.5 Theorem. *Let $(P_0, Q_0, P_1, Q_1, \ldots)$ be a commuting sequence of invertible operators of $B(X,X)$. Suppose that (N,p,X) and (N,q,X) are regular and that (P_n) and (Q_n) are each convergent in the norm of $B(X,X)$. Then (N,p,X) and (N,q,X) are consistent.*

Proof. Let A, B be elements of $B(X,X)$ such that $||P_n - A|| \to 0$ and $||Q_n - B|| \to 0$ $(n \to \infty)$. Since (N,p,X) is regular we have that $(||P_n^{-1}||)$ is bounded. Since $B(X,X)$ is a Banach algebra it follows that A is invertible; see for example Rickart [64], THEOREM (1.4.7). Likewise B is invertible. Now define

$$R_n = \sum_{k=0}^{n} Q_{n-k} P_k.$$

Then it is easy to check that, in the norm of $B(X,X)$,

$$(n+1)^{-1} R_n \to BA \quad (n \to \infty). \tag{7.22}$$

Since BA is invertible there exists m such that $(n+1)^{-1} R_n$ is invertible for all $n \geq m$, whence R_n is invertible for all $n \geq m$. Define a transformation, for $n \geq m$, by

$$M_n(Q,P,y) = \sum_{k=0}^{n} R_n^{-1} Q_{n-k} P_k y_k. \tag{7.23}$$

By (7.22) we have $(n+1)R_n^{-1} \to (BA)^{-1}$ $(n \to \infty)$, and so $||R_n^{-1}|| \leq H\,(n+1)^{-1}$ for some constant H, and all $n \geq m$. Since $(||P_n||)$ and $(||Q_n||)$ are bounded it follows readily that (7.23) defines a regular transformation.

Now

$$\sum_{k=0}^{n} Q_{n-k} P_k N_k^p(x) = \sum_{k=0}^{n} Q_{n-k} \sum_{r=0}^{k} p_{k-r} x_r = \sum_{r=0}^{n} (Q{*}p)_{n-r} x_r,$$

and

$$\sum_{k=o}^{n} P_{n-k} Q_k N_k^q(x) = \sum_{r=o}^{n} (P*q)_{n-r} x_r .$$

By the commuting property, $Q*p = P*q$ and $Q*P = P*Q$, whence for $n \geq m$,

$$M_n (Q,P,N^p(x)) = M_n(P,Q,N^q(x)).$$

Thus, by regularity of (7.23), if $x_n \to \ell$ (N,p,X) and $x_n \to \ell'$ (N,q,X), then $\ell = \ell'$, which proves the theorem.

We conclude with some remarks on the general problem of consistency for regular complex Nörlund means. The following is an open question:

7.6 Question. If (N,p) and (N,q) are any regular complex Nörlund means, must they be consistent?

From Theorem 7.5 we see that with some further restrictions on p and q the answer to Question 7.6 is in the affirmative.

It is interesting that the answer to the question is 'yes' provided the sequences p and q are real. Before commenting on the real case we remark that the positive case has long been known, and is due to Nörlund; see Theorem 7.7 below. A proof of Theorem 7.7 is in Hardy [19], p.65. Hardy's definition is in fact restricted to positive means.

7.7 Theorem. *Let (N,p) and (N,q) be regular positive Nörlund means, where (N,p) is called positive if and only if* $p_o > 0$ *and* $p_n \geq 0$ *for all* $n \geq 1$. *Then (N,p) implies (N,p*q), and (N,q) implies (N,p*q), whence (N,p) and (N,q) are consistent.*

In the following example it is shown that the result of Theorem 7.7 fails if one of the means is allowed to be conservative rather than regular, even with positivity.

7.8 Example. Define $p_0 = 1$, $p_n = 0$ $(n > 0)$, and $q_n = 2^n$. Then (N,p) and (N,q) are inconsistent.

Proof. Clearly (N,p) is regular, and (N,q) is conservative but not regular. If $x = (2, 0, 0, 0, \ldots)$ then $x_n \to 0$ (N,p) but $x_n \to 1$ (N,q).

Perhaps the 'best' result at present on consistency of Nörlund means is:

7.9 Theorem. Any two regular real Nörlund means are consistent.

By considering the relation between positive regular Nörlund means and a type of generalized Abel mean it was shown by Silverman and Tamarkin [69] (see also Hardy [19], p.65) that Theorem 7.9 held for positive Nörlund means. It was later observed by Jurkat and Peyerimhoff [24] that the proof of Silverman and Tamarkin was valid for the wider class of regular real Nörlund means, due essentially to the fact that if (N,q) is regular and real then Q_n is ultimately of constant sign.

Another proof of Theorem 7.9 was given by Thorpe [75] who showed that a 'modified' Nörlund mean $(\hat{N},p*Q)$ was implied by both of the regular real Nörlund means (N,p) and (N,q). Thus Thorpe's proof is in the spirit of Theorem 7.7, but with p*Q instead of p*q. The reason for having a modified mean is that

$$\sum_{k=0}^{n} (p*Q)_k \tag{7.24}$$

may be zero for some values of n, so that p*Q would not define a Nörlund mean (N,p*Q). However, when (N,p) and (N,q) are regular and real it follows, as shown by Thorpe, that the sum in (7.24) is ultimately non zero, and this is adequate for the purpose.

More recently, Kuttner [30] has shown that if (N,p) and (N,q) are regular real Nörlund means then there exists a regular real Nörlund mean (N,D) which is implied by both (N,p) and (N,q). Results connected with consistency of Nörlund means are also given by Kwee [32], [33].

BIBLIOGRAPHY

1. ALEXIEWICZ, A., On sequences of operations (I), Studia Math., 11 (1950), 1-30.

2. ALEXIEWICZ, A., On the two-norm convergence, Studia Math., 14 (1954), 49-56.

3. ALEXIEWICZ, A., and ORLICZ, W., Consistency theorems for Banach space analogues of Toeplitian methods of summability, Studia Math., 18 (1959) 199-210.

4. BANACH, S., Théorie des opérations linéaires, New York, 1955.

5. BENNETT, G., and KALTON, N.J., FK-spaces containing c_o, Duke Math. J., 39 (1972), 561-582.

6. BORWEIN, D., Linear functionals connected with strong Cesàro summability, J. London Math. Soc., 40 (1965), 628-634.

7. BRUDNO, A., Summation of bounded sequences by matrices, Mat. Sbornik, 16 (1945), 191-247.

8. CHANG, S.C., MACPHAIL, M.S., SYNDER, A.K., and WILANSKY, A., Consistency and replacability for conull matrices, Math. Zeit., 105 (1968), 2)8-212.

9. COOKE, R.G., Infinite matrices and sequence spaces, Macmillan and Co., London, 1949.

10. COPPING, J., Inclusion theorems for conservative summation methods, Nederl. Akad. Weten. Proc. Ser. A, 61 (1958), 485-499.

11. CRONE, L., A characterization of matrix operators on ℓ^2, Math. Zeit., 123 (1971), 315-317.

12. DAREVSKY, V., On intrinsically perfect methods of summation, Izv. Akad. Nauk. S.S.S.R. (Ser. Mat. N.S.), 10 (1946), 97-104.

13. DURAN, J.P., Infinite matrices and almost convergence, Math. Zeit., 128 (1972), 75-83.

14. DVORETZKY, A., and ROGERS, C.A., Absolute and unconditional convergence in normed linear spaces, Proc. Nat. Acad. Sci. (U.S.A.), 36 (1950), 192-197.

15. GARLING, D.J.H., The β- and γ-duality of sequence spaces, Proc. Camb. Phil. Soc., 63 (1967), 963-981.

16. HAHN, H., Über Folgen linearer Operationen, Monat. fur Math. und Phys., 32 (1922), 3-88.

17. HALMOS, P.R., Lectures on Ergodic Theory, Chelsea, New York, 1956.

18. HARDY, G.H., Theorems relating to the summability and convergence of slowly oscillating series, Proc. London Math. Soc. (2) 8 (1910), 301-320.

19. HARDY, G.H., Divergent Series, Oxford University Press, 1949.

20. HARDY, G.H., and LITTLEWOOD, J.E., Sur la série de Fourier d'une fonction à carré sommable, Comptes rendus, 156 (1913), 1307-1309.

21. HARDY, G.H., LITTLEWOOD, J.E., and POLYA, G., Inequalities, Cambridge University Press, 1967.

22. HARDY, G.H., and RIESZ, M., The general theory of Dirichlet's series. Cambridge, 1915.

23. JAMESON, G.J.O., Topology and normed spaces, Chapman and Hall, 1974.

24. JURKAT, W., and PEYERIMHOFF, A., The consistency of Nörlund and Hausdorff methods (Solution of a problem of E. Ullrich), Annals of Math., 62 (1955), 498-503.

25. KING, J.P., Almost summable sequences, Proc. Amer. Math. Soc., 17 (1966), 1219-1225.

26. KNOPP, K., and LORENTZ, G.G., Beiträge zur absoluten Limitierung, Archiv. der Math., 2 (1949), 10-16.

27. KÖTHE, G., Topological Vector Spaces I (English translation by D.J.H. Garling of Topologische Lineare Räume I, 1966), Springer-Verlag, 1969.

28. KÖTHE, G., and TOEPLITZ, O., Lineare Räume mit unendlichvielen Koordinaten und Ringe unendlicher Matrizen, J.f. reine u. angew. Math., 171 (1934), 193-226.

29. KUTTNER, B., Note on strong summability, J. London Math. Soc., 21 (1946), 118-122.

30. KUTTNER, B., Sequences which are summable by some regular Nörlund method, Aligarh Bull. of Math., 3-4 (1973-74), 1-14.

31. KUTTNER, B., A counter-example in summability theory, Math. Proc. Camb. Phil. Soc., 83 (1978), 353-355.

32. KWEE, B., The relation between Nörlund and generalized Abel summability, J. London Math. Soc., 38 (1963), 472-476.

33. KWEE, B., Some theorems on Nörlund summability, Proc. London Math. Soc., 14 (1964), 353-368.

34. LASCARIDES, C.G., A study of certain sequence spaces of Maddox and a generalization of a theorem of Iyer, Pacific J. Math., 38 (1971), 487-500.

35. LITTLEWOOD, J.E., The converse of Abel's theorem on power series, Proc. London Math. Soc. (2) 10 (1910/11), 434-448.

36. LORENTZ, G.G., A contribution to the theory of divergent sequences, Acta Math., 80 (1948), 167-190.

37. LORENTZ, G.G., and MACPHAIL, M.S., Unbounded operators and a theorem of A Robinson, Trans. Royal Soc. of Canada, XLVI (1952), 33-37.

38. MACPHAIL, M.S., Absolute and unconditional convergence, Bull. Amer. Math. Soc., 53 (1947), 121-123.

39. MADDOX, I.J., On Kuttner's theorem, J. London Math. Soc., 43 (1968), 285-290.

40. MADDOX, I.J., Elements of Functional Analysis, Cambridge University Press, 1970.

41. MADDOX, I.J., Kuttner's theorem for operators, Compos. Math., 29 (1974), 35-41.

42. MADDOX, I.J., Some general Tauberian theorems, J. London Math. Soc., 7 (1974), 645-650.

43. MADDOX, I.J., Schur's theorem for operators, Bull. Soc. Math. Grèce, 16 (1975), 18-21.

44. MADDOX, I.J., Tauberian estimates, J. London Math. Soc., 15 (1977), 143-146.

45. MADDOX, I.J., Matrix maps of bounded sequences in a Banach space, Proc. Amer. Math. Soc., 63 (1977), 82-86.

46. MADDOX, I.J., and WICKSTEAD, A.W., Crone's theorem for operators, Math. Colloq. Univ. Cap Town, 11 (1977), 33-45.

47. MADDOX, I.J., Consistency and Nörlund means, Math. Proc. Camb. Phil. Soc., 82 (1977), 107-109.

48. MADDOX, I.J., A new type of convergence, Math. Proc. Camb. Phil. Soc. 83 (1978), 61-64.

49. MADDOX, I.J., On strong almost convergence, Math. Proc. Camb. Phil. Soc., 85 (1979), 345-350.

50. MADDOX, I.J., Generalized Nörlund means and consistency theorems, Math. Proc. Camb. Phil. Soc. (to appear).

51. MARCINKIEWICZ, J., Sur la commabilité forte de séries de Fourier, J. London Math. Soc., 14 (1939), 162-168.

52. MAZUR, S., Über konvexe Mengen in linearen normierten Räumen, Studia Math., 4 (1933), 70-84.

53. MAZUR, S., and ORLICZ, W., Sur les méthodes linéaires de sommation, D.R. Acad. Sci. Paris, 196 (1933), 32-34.

54. MEHDI, M.R., Linear transformations between the Banach spaces L_p and ℓ_p with applications to absolute summability, Ph.D. thesis, University of London, 1959.

55. MELVIN-MELVIN, H., Generalized k-transformations in Banach spaces, Proc. London Math. Soc., 53 (1951), 83-108.

56. MEYER-KÖNIG, W., and TIETZ, H., On Tauberian conditions of type o, Bull. Amer. Math. Soc. 73 (1967), 926-927.

57. MEYER-KÖNIG, W., and TIETZ, H., Über die Limitierungsumkehrsatze vom Typ o, Studia Math. 31 (1968), 205-216.

58. MEYER-KÖNIG, W., and TIETZ, H., Über Umkehrbedingungen in der Limitierungstheorie, Arch. Math. (Brno) 5 (1969), 177-186.

59. NORTHCOTT, D.G., Abstract Tauberian theorems with applications to power series and Hilbert series, Duke Math. J., 14 (1947), 483-502.

60. ORLICZ, W., Linear operations in Saks spaces (II), Studia Math., 15 (1955), 1-25.

61. PETERSEN, G.M., Summability methods and bounded sequences, J. London Math. Soc., 31 (1956), 324-326.

62. PETERSEN, G.M., Regular matrix transformations, McGraw-Hill, 1966.

63. RAMANUJAN, M.S., Generalized Kojima-Toeplitz matrices in certain topological linear spaces, Math. Annalen, 159 (1965), 365-373.

64. RICKART, C.E., General theory of Banach algebras, Van Nostrand, 1960.

65. ROBERTSON, A.P., and ROBERTSON, W.J., Topological Vector Spaces, Cambridge University Press, 1964.

66. ROBINSON, A., On functional transformations and summability, Proc. London Math. Soc., 52 (1950), 132-160.

67. RUCKLE, W.H., The bounded consistency theorem, Amer. Math. Monthly, 86, No. 7 (1979), 566-571.

68. SCHUR, I., Über lineare Transformationen in der Theorie der unendlichen Reihen, J.f. reine u. angew. Math., 151 (1921), 79-111.

69. SILVERMAN, L.L., and TAMARKIN, J.D., On the generalization of Abel's theorem for certain definitions of summability, Math. Zeit., 29 (1929), 161-170.

70. STIEGLITZ, M., Fastkonvergenz und umfassendere durch Matrizenfolgen erklärte Konvergenzbegriffe. Habilitationsschrift. Universität Stuttgart, 1971.

71. STIEGLITZ, M., Eine Verallgemeinerung des Begriffs der Fastkonvergenz, Math. Japonicae, 18 (1973), 53-70.

72. STIEGLITZ, M., and TIETZ, H., Matrixtransformationen von Folgenräumen, Eine Ergebnisübersicht, Math. Zeit., 154 (1977), 1-16.

73. TAUBER, A., Ein Satz aus der Theorie der unendlichen Reihen, Monatshefte für Math. und Phy. 8 (1897), 273-277.

74. THORP, B.L.D., Sequential-evaluation convergence, J. London Math. Soc., 44 (1969), 201-209.

75. THORPE, B., An inclusion theorem and consistency of real regular Nörlund methods of summability, J. London Math. Soc., 5 (1972), 519-525.

76. TOEPLITZ, E., Über allgemeine lineare Mittelbildungen, Prace Mat. Fiz., 22 (1911), 113-119.

77. WIENER, N., Tauberian theorems, Annals of Mathematics. 33 (1932), 1-100.

78. WIENER, N., The Fourier integral and certain of its applications. Cambridge, 1933.

79. WILANSKY, A., Functional Analysis, Blaisdell Publ. Co., New York, 1964.

80. WILANSKY, A., Topological divisors of zero and Tauberian theorems, Trans Amer. Math. Soc., 113 (1964), 240-251.

81. WOOD, B., On ℓ-ℓ summability, Proc. Amer. Math. Soc., 25 (1969), 433-436.

82. ZELLER, K., Allgemeine Eigenschaften von Limitierungsverfahren die auf Matrix transformationen beruhen. Wissenschaftliche Abhandlung, 1949.

83. ZELLER, K., Allgemeine Eigenschaften von Limitierungsverfahren, Math. Zeit., 53 (1951), 463-487.

84. ZELLER, K., Verallgemeinerte Matrix transformationen, Math. Zeit., 56 (1952), 18-20.

85. ZELLER, K., Merkwürdigkeites bei Matrixverfahren; Einfolgenverfahren, Arch. Math. (Basel), 4 (1953), 1-5.

86. ZYGMUND, A., Trigonometric series, Vols. I and II, Cambridge University Press, 1959.

LIST OF SYMBOLS